AF700184

Math Mammoth Geometry 1

By Maria Miller

ISBN 978-1-954379-43-5

2024 EDITION

Contents

Introduction

Math Mammoth Geometry 1 covers all the typical geometry topics for grades 4-5: angles, triangles, quadrilaterals, other polygons, area, perimeter and volume.

Note: Any problem marked with “ ” means the exercise should be done in a notebook or on blank paper.

The book starts out by reviewing the third grade concepts of area and the perimeter of rectangles. Students also apply these concepts in various problems, including problems where they write simple equations and a problem where they explore all possible perimeters for a given area.

Note: Students will need a ruler and a protractor throughout the book.

The first part of the book covers 4th grade topics, with a focus on angles. Students learn about lines, rays, and angles; and about acute, right, obtuse, and straight angles. They learn how to measure and draw angles with a protractor. We also study angle problems where students write simple equations. The lesson *Estimating Angles* has an optional section on turning in an angle, which can be challenging, so feel free to omit it if you wish.

The lesson *Parallel and Perpendicular Lines* also ties in with the topic of angles, because two lines are perpendicular if they form a right angle. After that, we study parallelograms and other quadrilaterals in more detail, paying attention to their angles and lengths of sides.

We also study triangles and classify them according to their angles (acute, obtuse, or right triangles). The last (and easy) topic in the first part of the book is line symmetry.

The latter half of the book covers 5th grade topics, starting with the lesson “Review: Angles”. The focus of this portion is on two topics: classifying two-dimensional shapes, and volume.

The lesson “Polygons” covers the concept of a polygon and the names of several common ones. Students classify figures into polygons and non-polygons, and also into regular polygons versus non-regular polygons.

The next topic is classifying quadrilaterals. The focus is on understanding the classification, and understanding that attributes defining a certain quadrilateral also belong to all the “children” (subcategories) of that type of quadrilateral. For example, squares are also rhombi, because they have four congruent sides (the defining attribute of a rhombus).

A possible confusion point is the definition of a trapezoid. There exist two possible definitions:

- (Exclusive definition:) A trapezoid has exactly one pair of parallel sides.
- (Inclusive definition:) A trapezoid has at least one pair of parallel sides.

Both definitions are legitimate, but lead to different analysis when classifying quadrilaterals. Under the exclusive definition, a parallelogram is not a trapezoid, but under the inclusive definition, it is. Most college-bound textbooks favor the *inclusive* definition, and that is what is used in this text, also.

Then we study the classification of triangles. Students are now able to classify triangles both in terms of their sides and also in terms of their angles.

The second focus topic of this latter half is volume. Students learn that a cube with the side length of 1 unit, called a “unit cube,” is said to have “one cubic unit” of volume, and can be used to measure volume. They find the volume of right rectangular prisms by “packing” them with unit cubes and by using formulas. They recognize volume as additive and solve both geometric and real-word problems involving volume.

The book includes three optional lessons listed in the end: area and perimeter problems, star polygons, and circles. Use them as time allows. The lesson on area and perimeter can be important for those students who tend to forget these concepts. The lesson on star polygons is intended as a fun artistic topic. The lesson on circles involves the usage of a compass, which may be hard for some children at this age. Those who can master it will probably find the exercises involving multiple circles fascinating.

Geometry is a hands-on subject, and thus the lessons include quite a few drawing exercises which can be done on blank paper, in a notebook, or in the worktext (for most). I hope the students enjoy the hands-on experience. Besides, these drawing exercises are a good means of achieving a true conceptual understanding of geometry concepts.

Please stress to the student to always use a ruler and other proper tools, such as a protractor or a triangular ruler, so the drawings will be as accurate as possible. Some exercises may mention to only sketch something, in which case it is okay to not use any drawing tools.

Geometry is full of strange-sounding words. I suggest that student(s) keep a geometry notebook, where they draw picture(s) and text to explain every new concept or term. This will help them to remember those terms. They can also do the drawing exercises in the notebook. Encourage the students to be creative so that the notebook becomes their own special work. You can even give them credit for it.

Please see also the videos at **https://www.mathmammoth.com/videos/geometry_1**. Most of them match the topics in this book, and can be used to provide additional instruction.

I wish you success in teaching math!
Maria Miller, the author

Helpful Resources and Games on the Internet

We have compiled a list of external Internet resources that match the topics in this book. This list of links includes web pages that offer:

- **online practice** for concepts;
- online **games**, or occasionally, printable games;
- **animations** and interactive **illustrations** of math concepts;
- **articles** that teach a math concept.

We heartily recommend you take a look at the list. Many of our customers love using these resources to supplement the bookwork. You can use the resources as you see fit for extra practice, to illustrate a concept better, and even just for some fun. Enjoy!

https://l.mathmammoth.com/blue/geometry1

SCAN ME

Geometry Vocabulary Reference Sheet

I encourage you to draw pictures to illustrate the terms, or even make your own geometry notebook!

Two lines are **perpendicular** if they form a right angle.

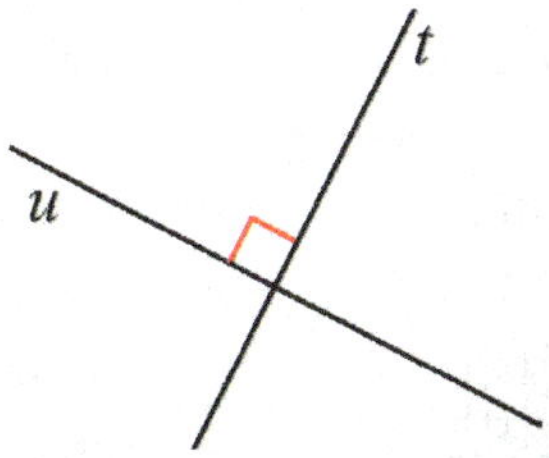

An **angle** consists of two rays that start at the same point, called vertex. The two rays form the sides of the angle.

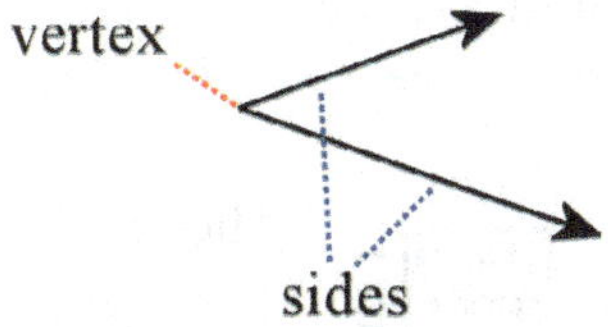

- A **polygon** is a flat, two-dimensional figure that consists of line segments, and is closed.
- A **regular polygon** is one with congruent sides and angles.
- A **vertex** is a "corner" of a polygon.
- A **diagonal** is a line segment drawn from one vertex of a polygon to another.

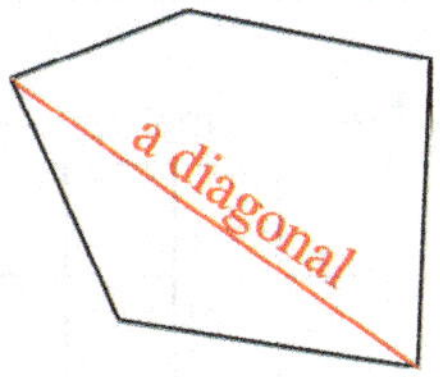

- A **quadrilateral** – a polygon with *four* sides
- A **pentagon** – a polygon with *five* sides.
- A **hexagon** – a polygon with *six* sides.
- A **heptagon** – a polygon with *seven* sides.
- An **octagon** – a polygon with *eight* sides.

- A **right triangle** is a triangle with one right angle.
- An **obtuse triangle** is a triangle with one obtuse angle.
- An **acute triangle** is a triangle with all three angles acute.

- An **equilateral triangle** is a triangle with three congruent sides.
- An **isosceles triangle** is a triangle with two congruent sides.
- A **scalene triangle** is a triangle where none of the sides are congruent.

- A **trapezoid** is a quadrilateral with at least one pair of parallel sides.
- A **parallelogram** is a quadrilateral with two pairs of parallel sides.
- A **rhombus** is a parallelogram with four congruent sides.
- A **kite** is a quadrilateral that has two pairs of congruent sides, and the congruent sides are adjacent (neighboring each other).
- A **rectangle** is a quadrilateral with four right angles.
- A **square** is a rectangle with four congruent sides.
- A **scalene quadrilateral** has no congruent sides.

- A **rectangular prism** is a box-shaped solid (three-dimensional shape) with edges that meet at right angles.

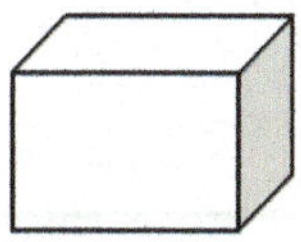

Review: Area of Rectangles

Area is always measured in **squares of some size**. We use the superscript "2" with a unit of length to indicate the "squaring." For example, 120 cm^2 means 120 square centimeters.

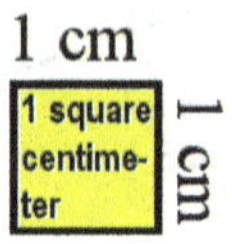

The area of this square is 1 square centimeter, or 1 cm^2.

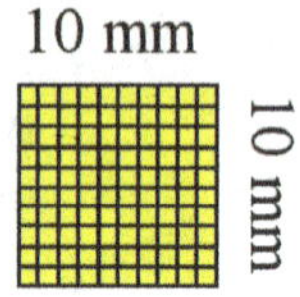

Each tiny square has an area of 1 square millimeter, or 1 mm^2. The area of the whole square is 10 mm × 10 mm = 100 mm^2.

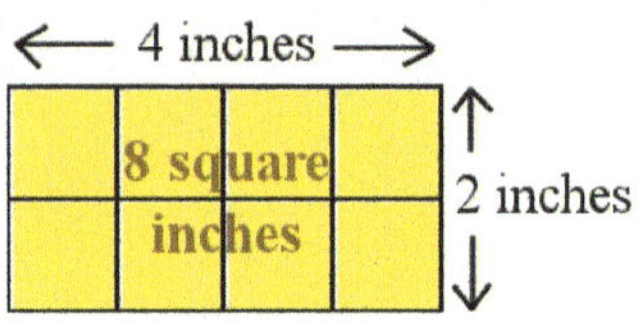

You can use multiplication to find how many square inches this rectangle covers: 2 in × 4 in = 8 in^2.

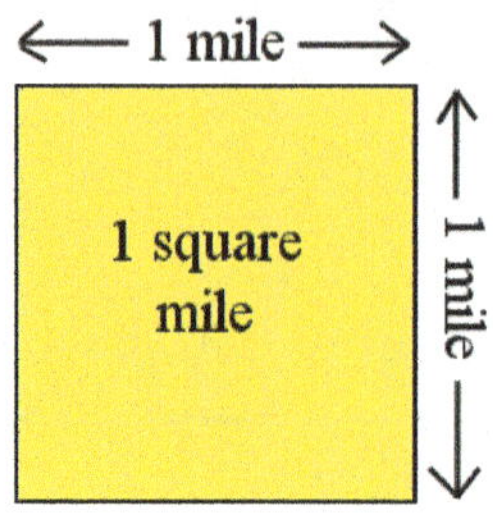

The area of this square is 1 mi × 1 mi = 1 square mile, or 1 mi^2.

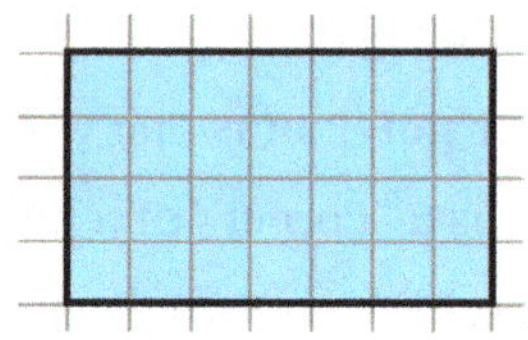

If no particular unit of length is given for the sides, we just use the word "unit" for the lengths. The area is then 7 units × 4 units = 28 *square units.*

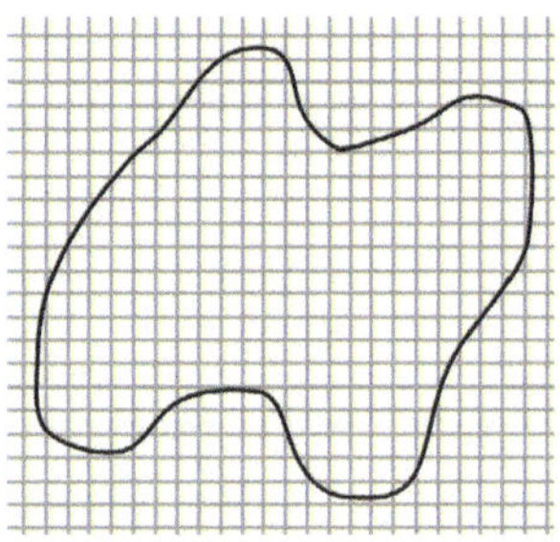

If the figure is some other shape than a rectangle, we will still use little squares to measure its area. It is just more difficult to find out how many little squares it covers, and we may have to use partial (fractional) squares as well.

1. Write a multiplication to calculate the area of these rectangles. <u>Include the units!</u>

a.

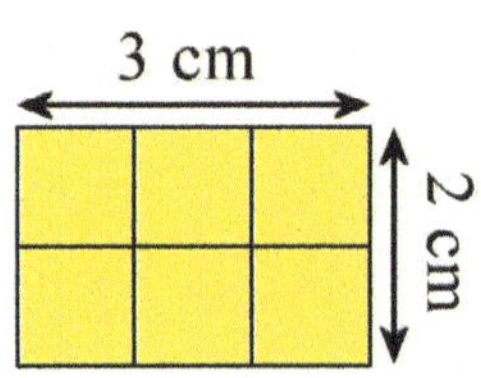

A = ________ × ________

= ____________________

b.

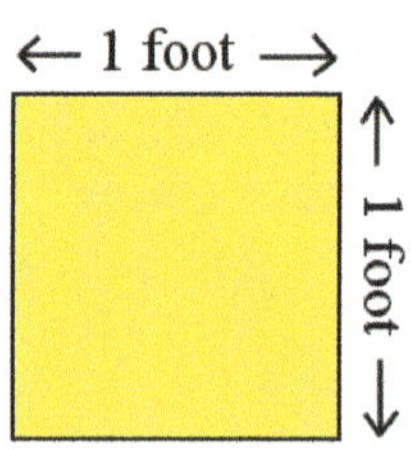

A = ________ × ________

= ____________________

c.

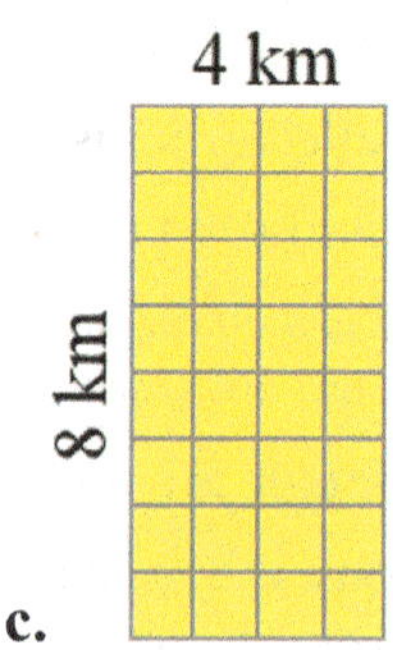

A = ________ × ________

= ____________________

2. Find the missing measurements.

a. 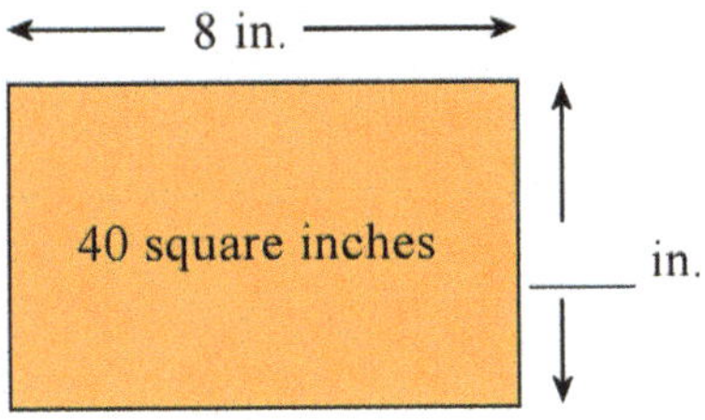

b. 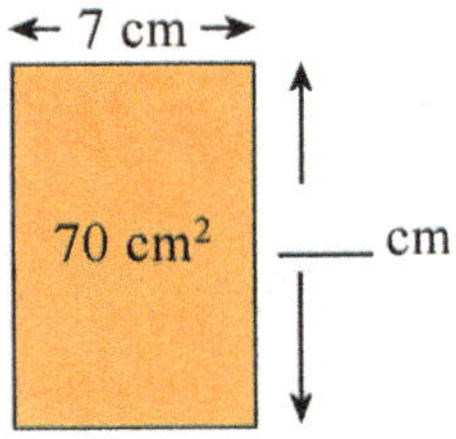

c. 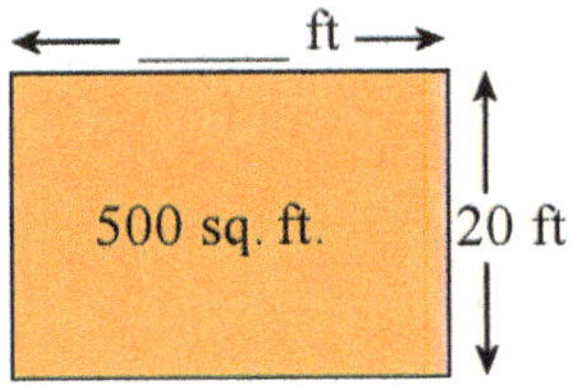

3. Write an equation (a multiplication with an unknown) for the area. Then solve.

a. 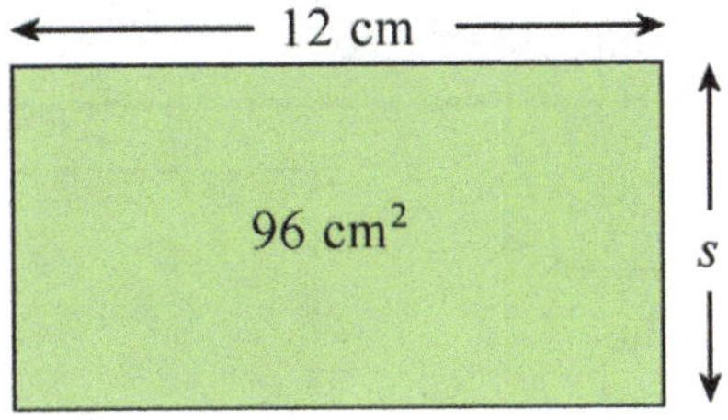

A = ______ × ______ = ________

s = __________

b. 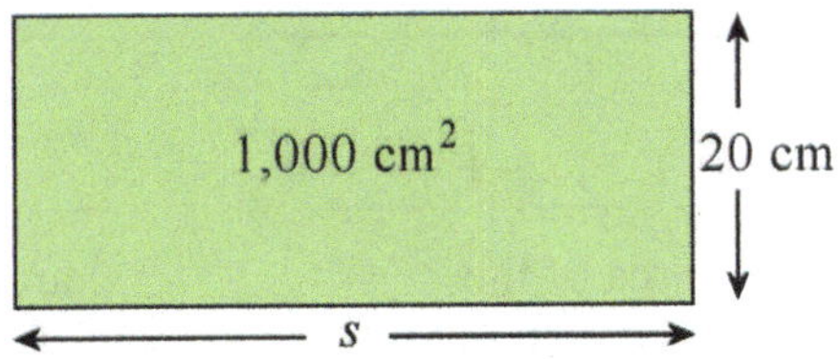

A = ______ × ______ = ________

s = __________

4. Write an equation (a multiplication with an unknown) for the area. Then solve.

a. The area of a rectangle is 45 m^2, and its one side measures 9 m.
How long is the other side (s)?

______ × ______ = ________; s = __________

b. The area of a rectangular field is 1,800 ft^2, and its one side measures 60 ft.
How long is the other side (s)?

______ × ______ = ________; s = __________

5. This is a plan for a doghouse. In the grid, the side of each little square is 6 inches.

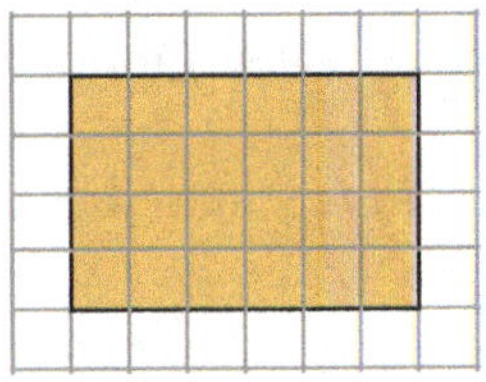

a. What is the area of the doghouse in square inches?

b. Figure out the lengths of sides in feet.
Then calculate the area in square feet.

If we think of this rectangle as two rectangles, we can write its area as $3 \times 3 \ + \ 3 \times 5$.

But if we think of it as one rectangle with sides 3 units and (3 + 5) units, then we can write its area as $3 \times (3 + 5)$.

Those two are of course equal. So, thinking of it as one rectangle or two rectangles, we get:

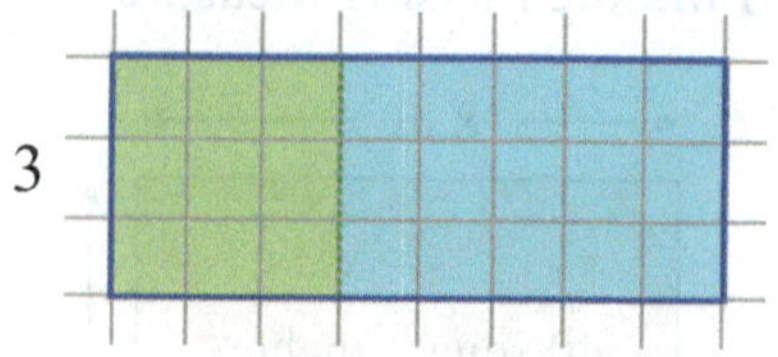

$3 \times (3 + 5)$	=	3×3	+	3×5
area of the whole rectangle		area of the first part		area of the second part

6. Write a number sentence for the total area, thinking of one rectangle or two.

a. ___ × (___ + ___) = ___ × ___ + ___ × ___

area of the whole rectangle — area of the first part — area of the second part

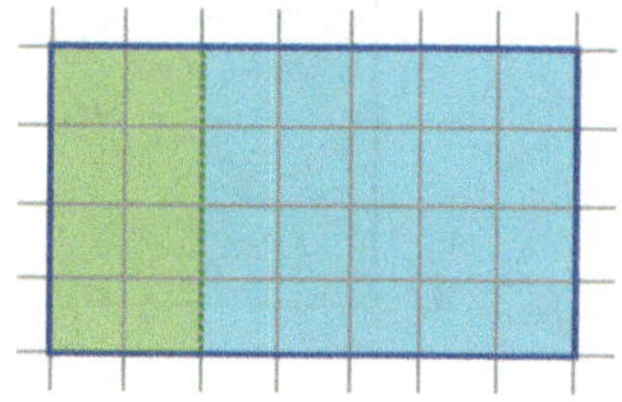

b. ___ × (___ + ___) = ___ × ___ + ___ × ___

area of the whole rectangle — area of the first part — area of the second part

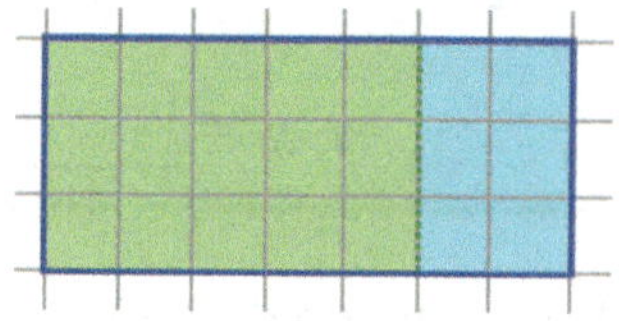

7. This is a plan for a two-part clubhouse.

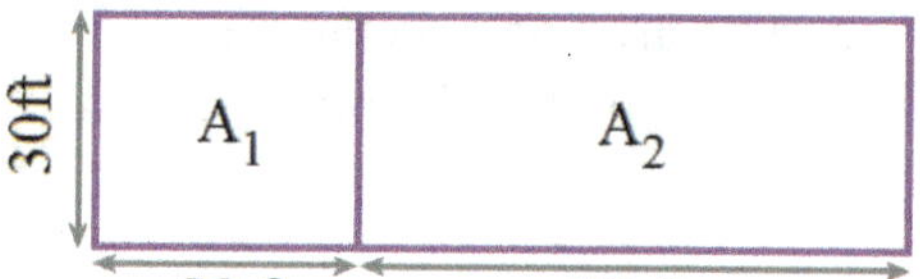

a. Write a multiplication and addition sentence for the total area, thinking of one rectangle or two. Solve it.

A = ________ × (________ + ________)

= ________ × ________ + ________ × ________

= ______________________________________

b. How much larger in area is the part A_2 than the part A_1?

8. Answer. Optionally, you can sketch these squares on blank paper to help you.

a. A square has an area of 4 square inches. How long is its side? ______________

b. A square has an area of 9 cm^2. How long is its side? ______________

c. A square has an area of 1 square foot. How long is its side? ______________

Problem Solving: Area of Rectangles

1. Write a number sentence for the area of each figure. Use multiplication, addition, and/or subtraction.

a.

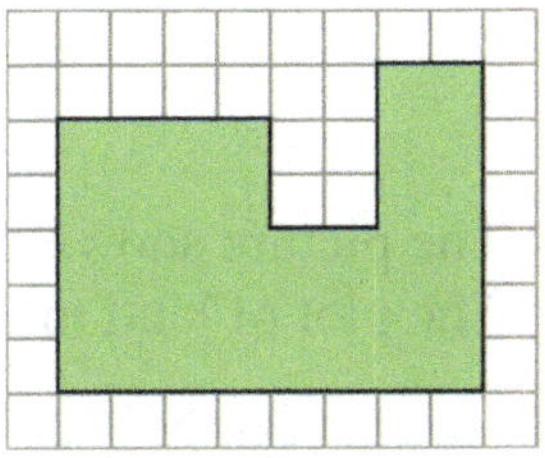

b.

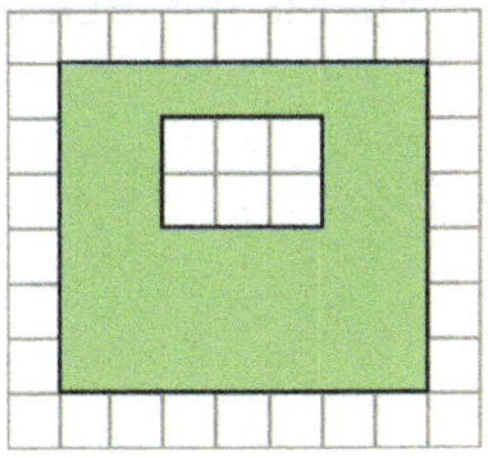

2. The second floor of a doll house has three rooms. A room at the end measures 20 cm by 25 cm, the middle room is 40 cm by 25 cm, and the third one is 20 cm by 25 cm.

   a. Draw a sketch of the second floor.

   b. Find its area.

3. Draw an L-shape with an area of at least 20 square units. Write a number sentence for its area.

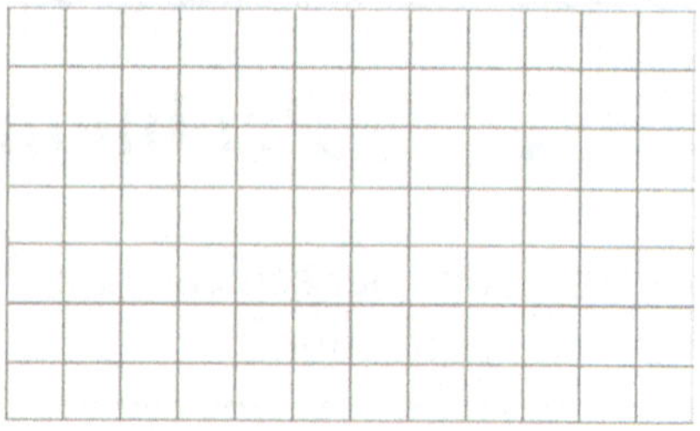

Area = ________________________________

____

4. The picture shows a rectangular piece of land with a house in the middle. The plot of land measures 48 ft by 48 ft, and the house is 30 ft by 20 ft.

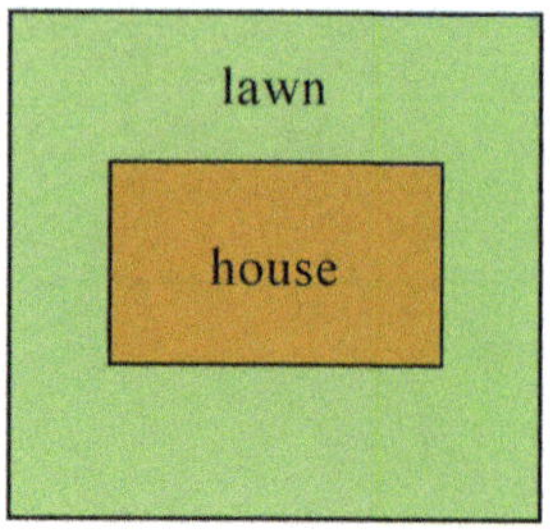

What is the area of the lawn?

5. Find the areas of the shaded figures.

a.

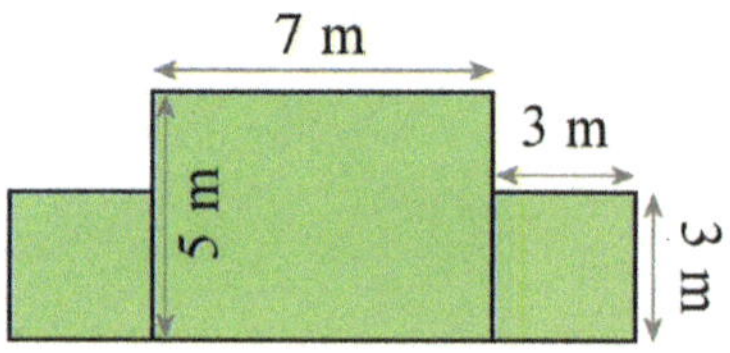

b.

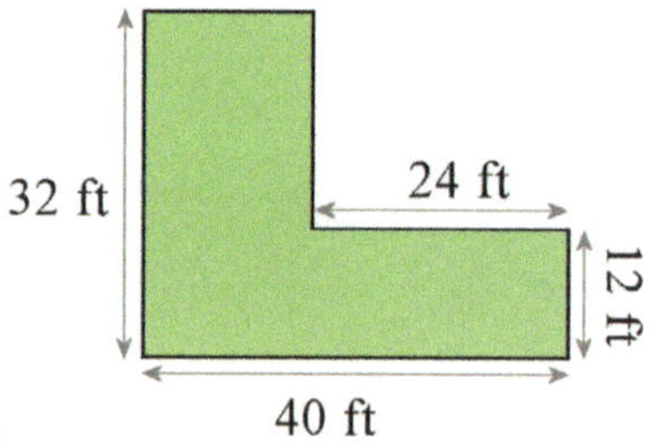

Review: Area and Perimeter

If you confuse perimeter and area, just remember this:

- AREA has to do with coverage. We think of covering the shape with squares. Area is measured in some kind of squares, such as square centimeters, square inches, square feet, square meters, or just square units.
- PERIMETER has to do with "going all the way around." Perimeter is given in some unit of length, such as centimeters, meters, inches, or feet.

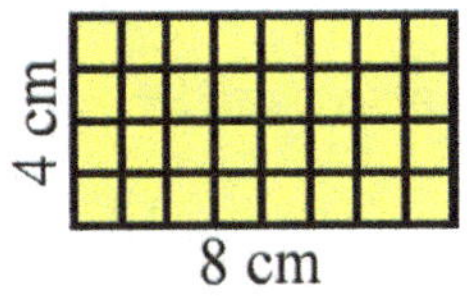

Area: 4 cm × 8 cm = 32 cm^2.

Perimeter:
4 cm + 8 cm + 4 cm + 8 cm = 24 cm

1. Which quantity—perimeter, area, or volume—fits each situation, if you need to find out...

   a. how much fence is needed to go around a yard?

   b. how much water fits into a bottle?

   c. what size of a carpet you need to cover the floor?

2. Find the area and perimeter of the rectangles. Don't forget the right kind of unit in your answer!

a. 12 in. wide, 4 in. tall

Perimeter = ____________________

Area = ____________________

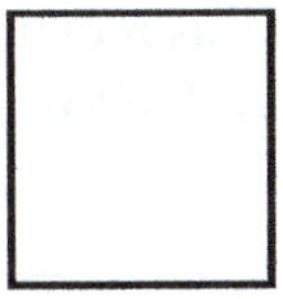

b. All sides are 50 cm.

Perimeter = ____________________

Area = ____________________

3. In this grid, each square measures 1 cm by 1 cm.
   a. What is the area and perimeter of the L-shape?

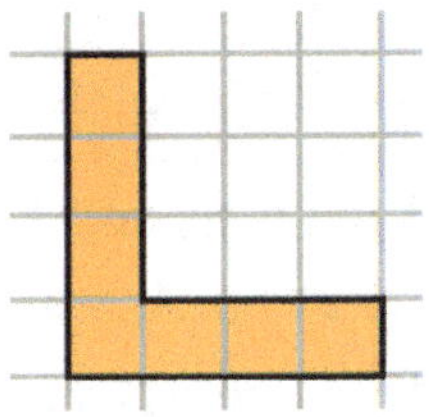

   Perimeter = ____________________

   Area = ____________________

   b. If each little square measured 2 cm by 2 cm, then what would the area and perimeter be?

   Perimeter = ____________________

   Area = ____________________
   (*Hint: Find the area of each little square.*)

The perimeter of a rectangle is 46 cm. Its one side is 7 cm. How long is the other side?

We can write an addition equation with an unknown:

7 + ? + 7 + ? = 46

We could also use x:

$7 + x + 7 + x = 46$

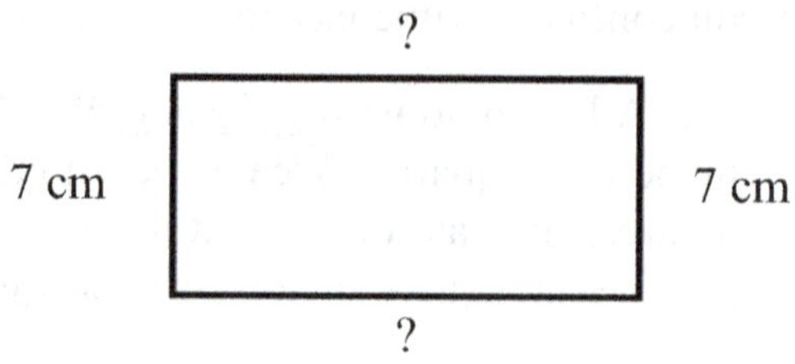

We could guess and check to solve it. But, there is also an easier way. Just think: the two sides, 7 and ?, form ***half*** of the perimeter. So, 7 + ? = 23.

Thinking either way, we can solve that ? = 16 cm.

4. Solve. Write a number sentence with an unknown for problems a, b, and c.

a. The perimeter of this rectangle is 32 ft. Its one side is 9 ft. How long is the other side?

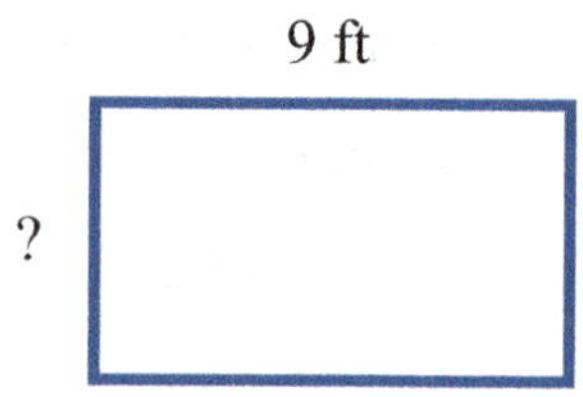

________ + ? + ________ + ? = ________

Solution: ? = ________

b. The perimeter of this rectangle is 44 cm. One side is 15 cm. How long is the other side?

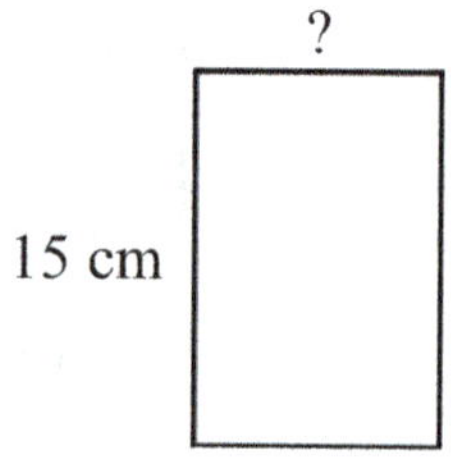

____

Solution: ____ = ________

c. A square has a perimeter of 24 m. How long is its side?

____

Solution: ____ = ________

What is its area? ________________

d. A square has an area of 25 cm^2. How long is its side?
This time, don't worry about writing a number sentence, but you may, if you are able.

____

____

What is its perimeter? ________________

5. Fill in. Draw a rectangle for each problem.

a. Sides 4 and _____ units; area 20 square units, perimeter ______ units.	**b.** Sides ____ and ____ units; area 15 square units, perimeter 16 units.	**c.** Sides 3 and _____ units; area _____ square units, perimeter 20 units.

6. The perimeter of one of these hexagons is 516 mm.
How long is its side?

Photo courtesy of
http://www.flickr.com/photos/aubergene

7. **a.** Write a number sentence for the area of this school building. Notice that two of the measurements are missing.

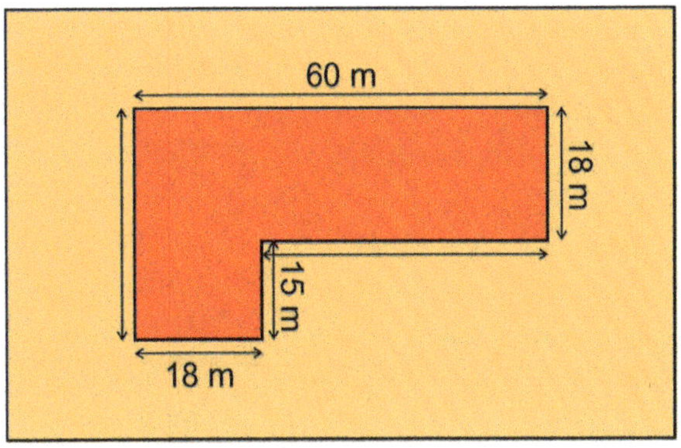

b. Calculate the area.

c. *(Challenge)* The outer yard measures 80 m by 35 m.
Find the area of the actual yard—that is, not including the building.

8. A farmer wants to build a sheep yard that is 400 square meters in area.

a. Suggest three different rectangular shapes for the sheep yard. Give their width and length. For example, one possibility would be 100 meters long and 4 meters wide, but that is very narrow!

	Length	Width	Area	Fencing needed
Sheep yard 1			400 m^2	
Sheep yard 2			400 m^2	
Sheep yard 3			400 m^2	

b. For each possible sheep yard, calculate how much fence he needs to go around it.

9. You want to build a rectangular dog pen so that 30 m of fencing is enough to go around it. The pen should be at least 50 m^2 in area (it can be larger). Suggest a length and width for the dog pen. *(Hint: Draw sketches.)*

Puzzle Corner

The outer length and width of a picture frame are 19 cm and 14 cm. The frame is 2 cm wide. What is the area of the *inside*?

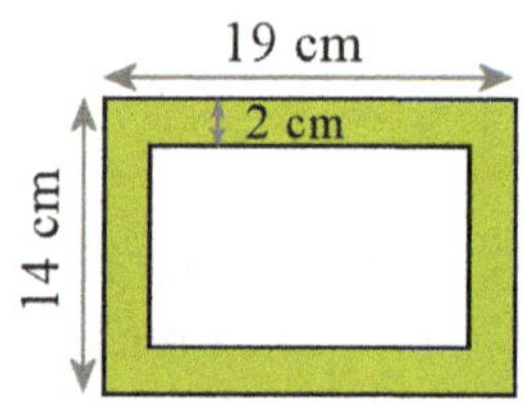

Lines, Rays, and Angles

This is point A. (Points are named with capital letters.) 	This is a **line segment**. We write this as line segment AB or line segment $\overline{AB}$. 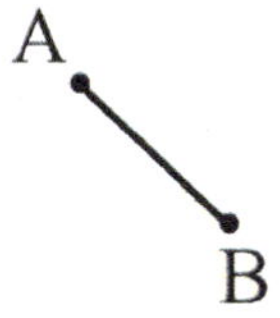

A line has no beginning point or end point. Imagine it continuing indefinitely in both directions. We can illustrate that by little arrows on both ends.

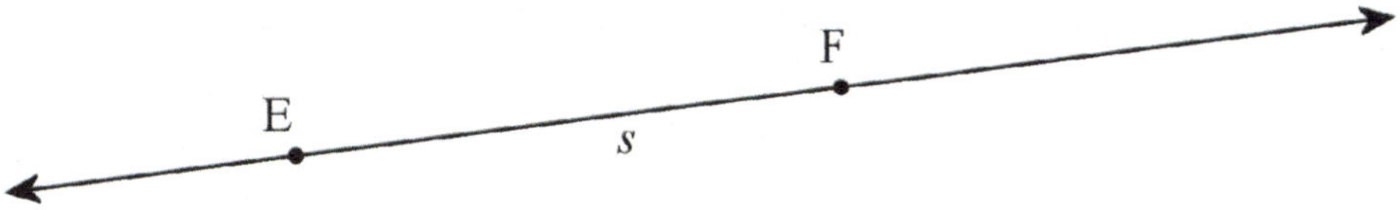

A line is named using two points on it. This is line EF or line $\overleftrightarrow{EF}$ (note the arrows on both ends). Or, we can name a line using a lowercase letter: this is line *s*.

A ray starts out at a point but continues on indefinitely, without ending. We can show that by drawing an arrow at one end of the ray. Think of the sun's rays!

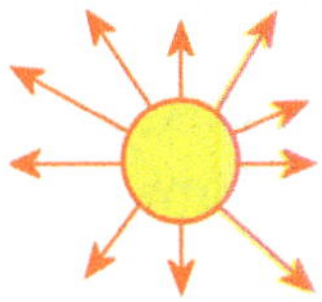

A ray is named using its starting point and one other point on the ray: this is ray QP or ray $\overrightarrow{QP}$ (note the one arrow). Or, we can name a ray using a lowercase letter: this is ray *r*.

1. Write the name if each figure is a line, ray, or a line segment.

a. ________________

b. ________________

c. ________________

2. **a.** Draw the ray BD.

   b. Draw the line AB.

   c. Draw the line segment ED.

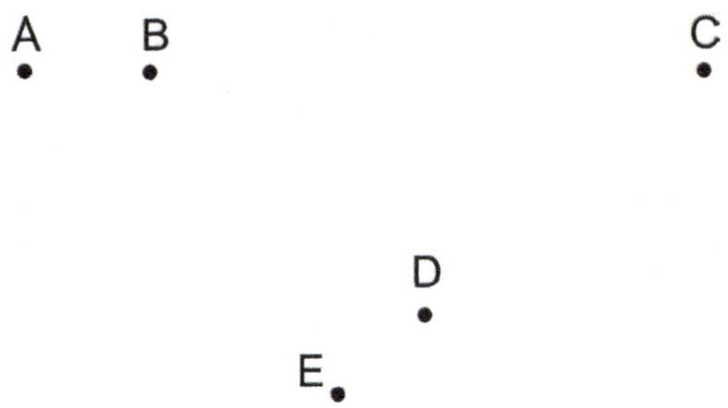

What is an angle? Many people think that an angle is some kind of slanted line. However, in geometry, **an angle** is made up of **TWO RAYS that have the same beginning point**.

That point is called the **vertex**, and the two rays are called the **sides** of the angle.

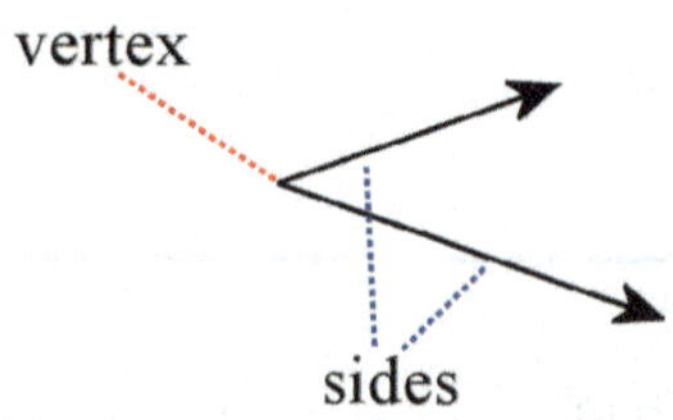

To name an angle, we use three points, listing the vertex in the middle. This is angle DEF or ∠DEF. We can use the symbol ∠ for angle.

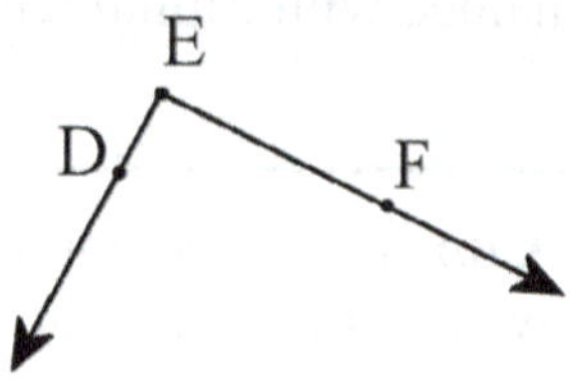

3. Name the angle.

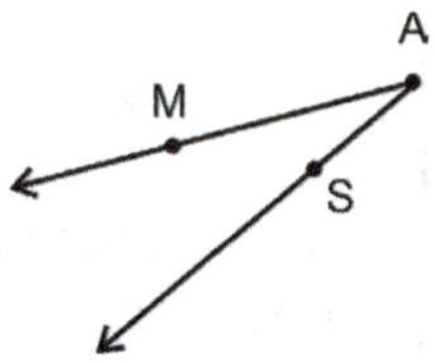

4. **a.** Find the angle formed by the rays DE and DF.
How do we name it?

b. Find the angle formed by the rays CA and CE.
How do we name it?

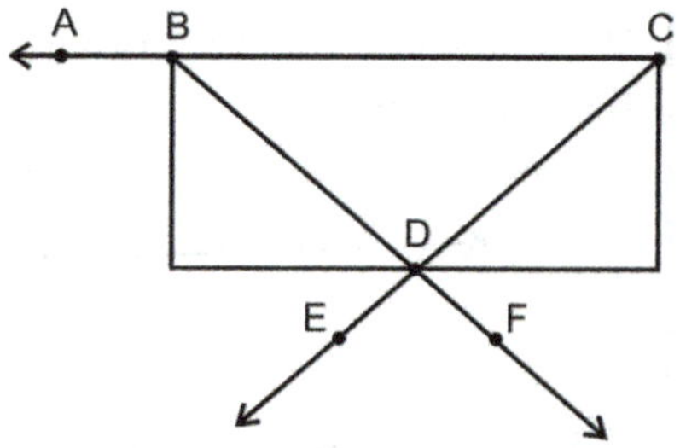

5. **a.** Draw two points, D and E. Then draw line DE.

b. Draw point *Q not* on the line.

c. Draw rays DQ and EQ.

d. Find angles EDQ and DEQ in your drawing.

Angles "open up"

Take two pencils to illustrate the two rays which make the two sides of an angle. Set them side by side to show a zero angle! 

Then start "opening up" the angle. Keep one pencil stationary while you rotate the other. As the angle opens up, an imaginary arc of a circle is drawn.

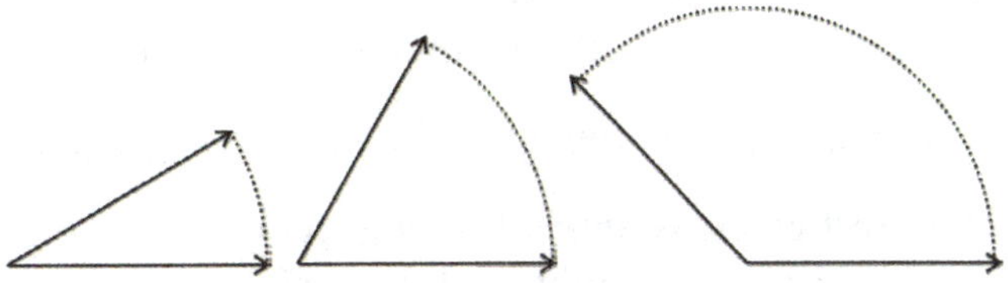

Telling the size of angles

Which of these two angles is bigger?

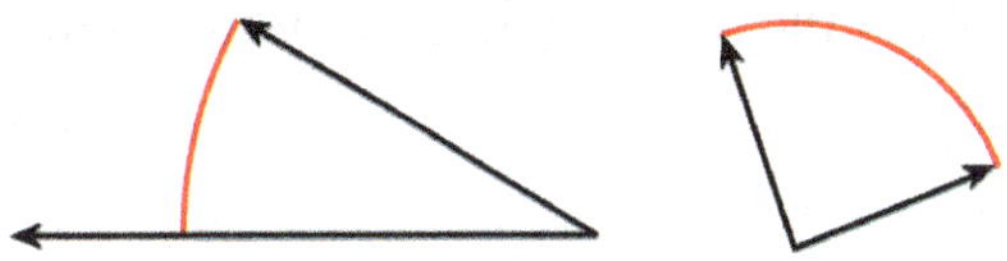

Do not look at how LONG the sides of the angle are. Remember, the sides of an angle are rays, and rays go on indefinitely (even if they are drawn short).

Instead, look at <u>how much each angle has *opened*</u>, or how big a part of a circle the sides have drawn. The second angle (on the right) is bigger.

The **size of an angle** is determined by **how much it has opened as compared to the whole circle**.

Many times the arrows are omitted from the rays, and the arc of the circle is drawn as a tiny arc near the vertex. Even that is not necessary.

Which of these is the bigger angle?
Again, the second one is bigger.

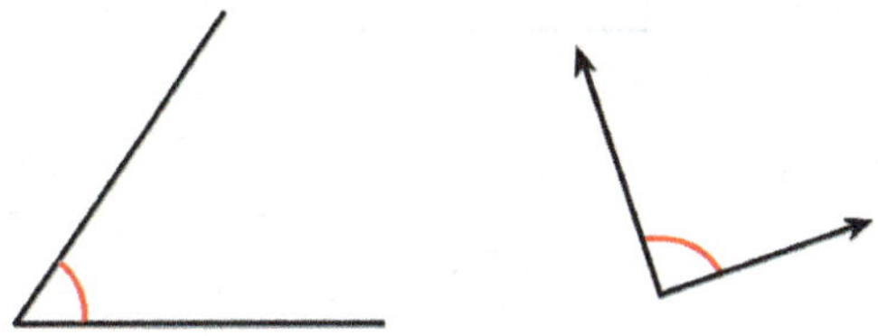

6. Which angle is bigger? You can use pencils to help!

a.  OR 

b.  OR 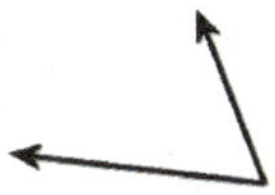

c. 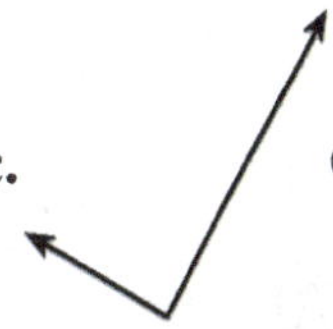 OR 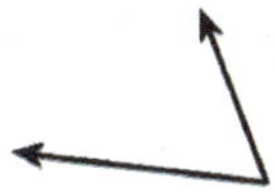

d. 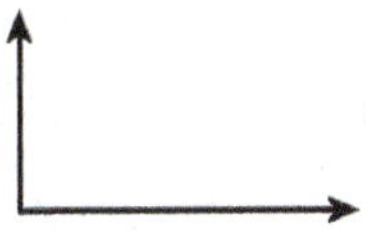 OR 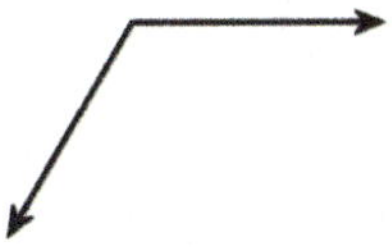

e. 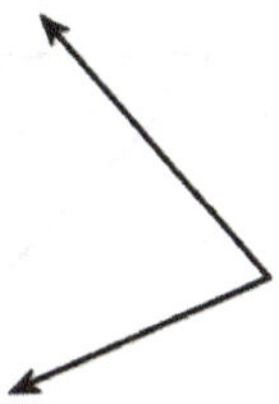 OR 

f. 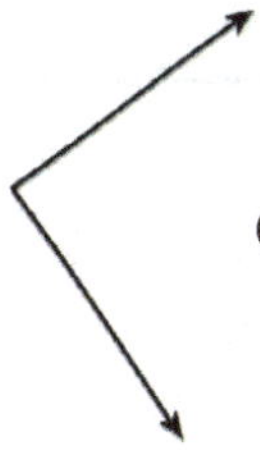 OR 

New Terms & Symbols	• *line segment* • *line*	• *ray* • angle

Note: In the next lesson, students will need a *protractor* to measure angles.

Measuring Angles

Remember how one side of the angle traces out a circular arc°. We use that circle to measure how big the angle is. We look at how much the angle has "opened" as compared to the full circle.

Angles are measured in **degrees**. The symbol for degrees is a small raised circle: °.

- **The full circle is 360° (360 degrees).**
- A half circle (a straight angle) is 180°.
- A quarter of a circle (a right angle) is 90°.

Show the angles below using two pencils. Try to "see" the circle that is traced in the air.

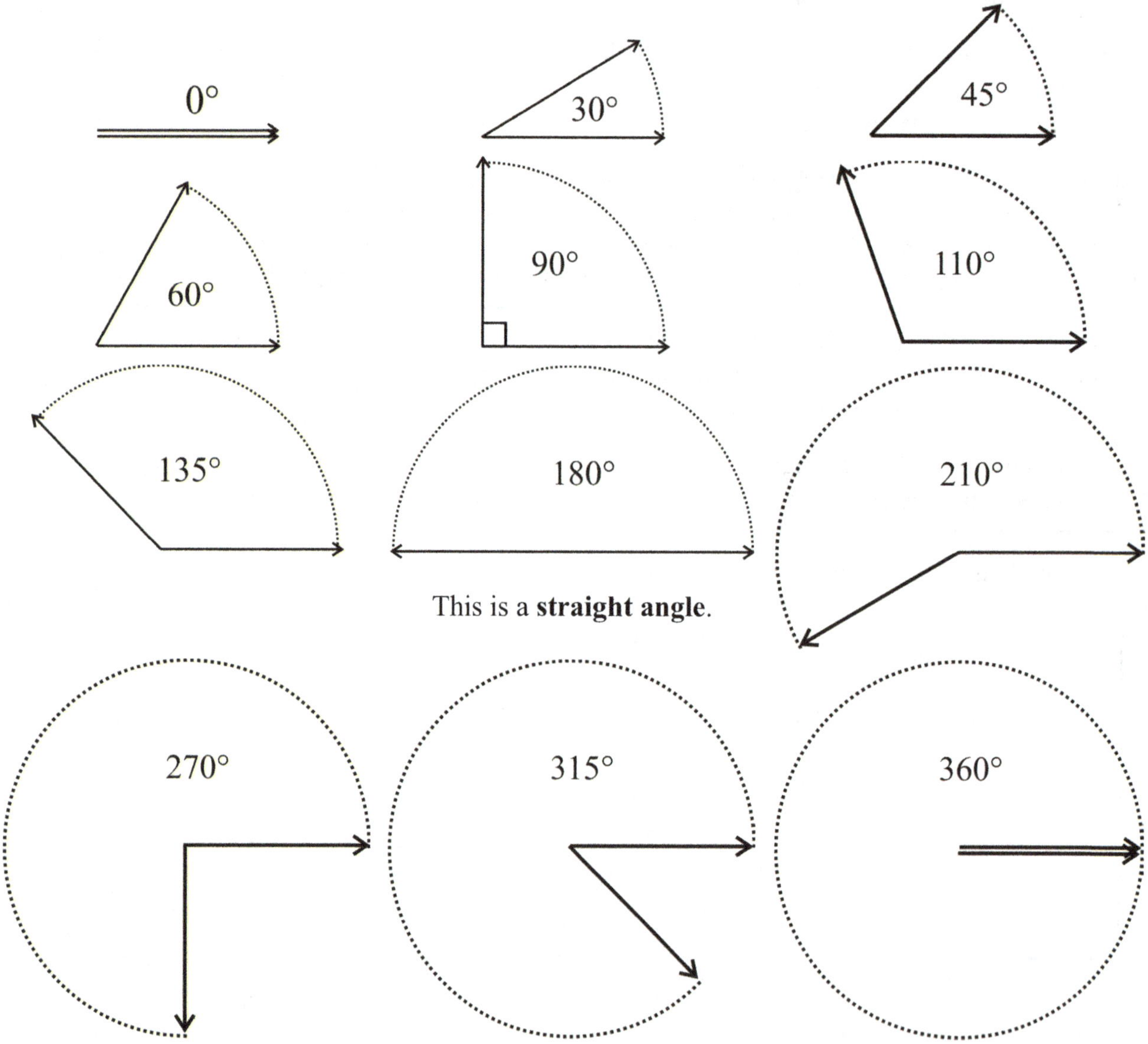

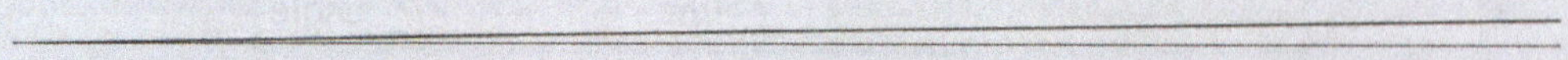

This is a 1-degree angle — it is 1/360 part of the full circle!

How to use a protractor to measure angles

1. Place the midpoint of the protractor on the vertex of the angle.
2. Line up one side of the angle with the zero line of the protractor (where you see the number 0).
3. Read the degrees where the other side crosses the number scale.

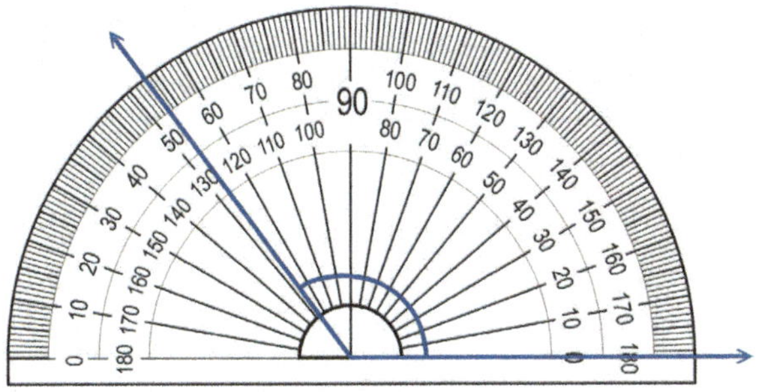

an obtuse angle; 127°

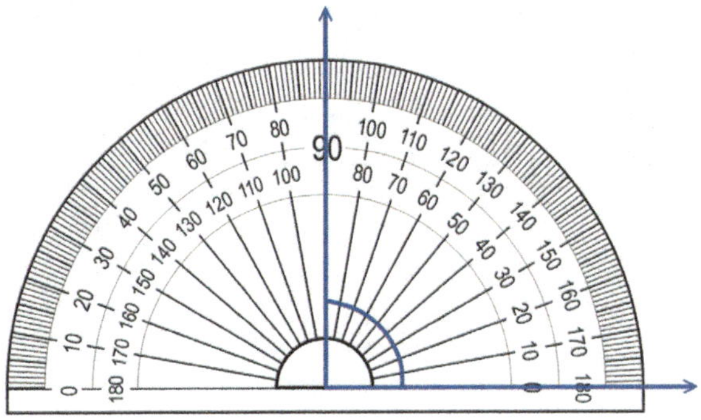

a right angle; 90°

Make sure you read from the right set of numbers. A protractor has two sets of numbers: one set goes from 0 to 180, and the other set from 180 to 0. Which one you read depends on how you place the protractor in relation to the angle. One of the sides of the angle is lined up with one of the zero lines of the protractor. You read the set of numbers that starts with *that* zero.

In the examples above we lined up one of the sides of the angle with the zero of the lower set of numbers, so we need to read the lower set of numbers.

1. Measure the angles.

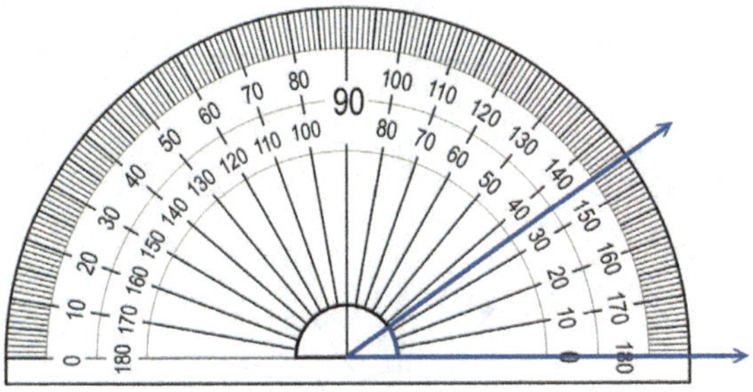

a. __________ °

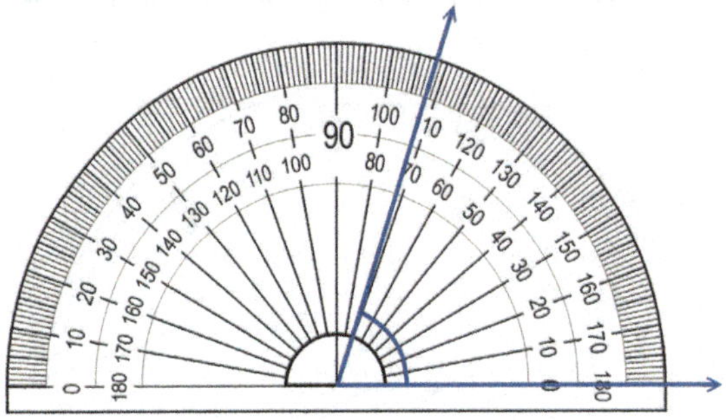

b. __________ °

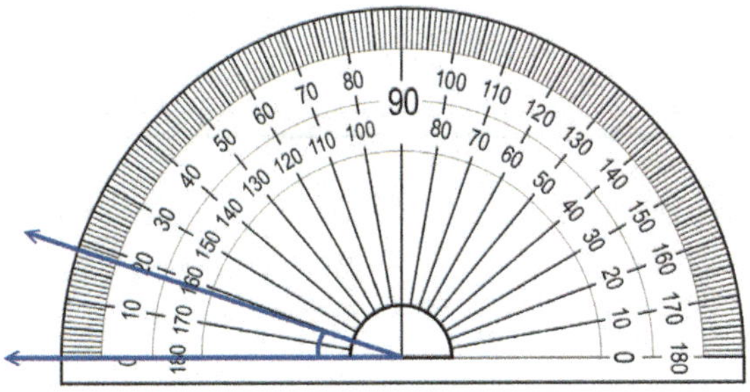

c. __________ °

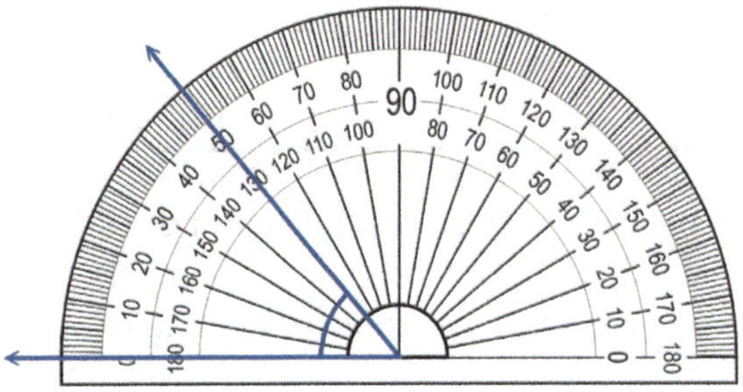

d. __________ °

2. Measure the angles.

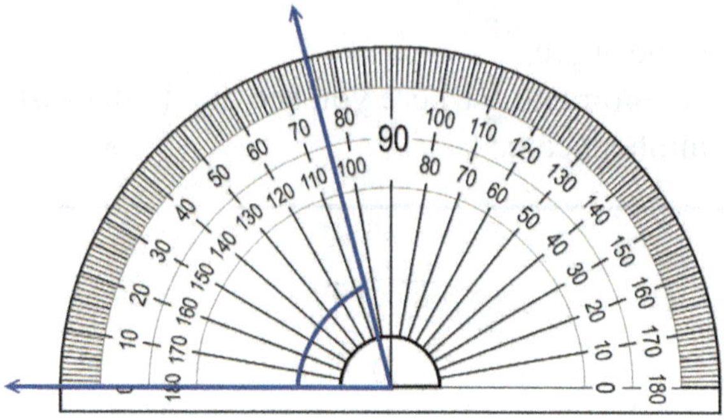

a. __________°

____

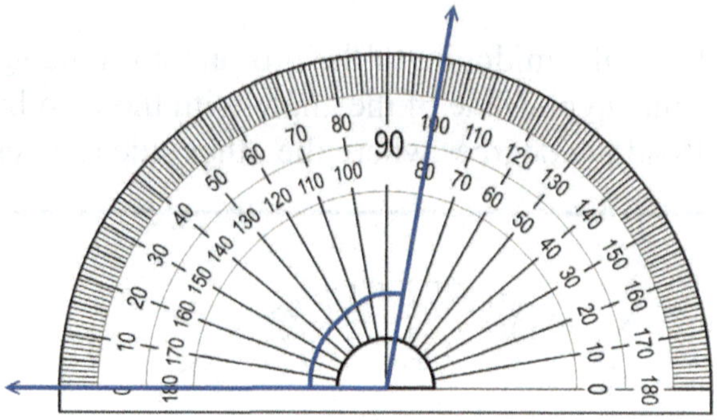

b. __________°

____

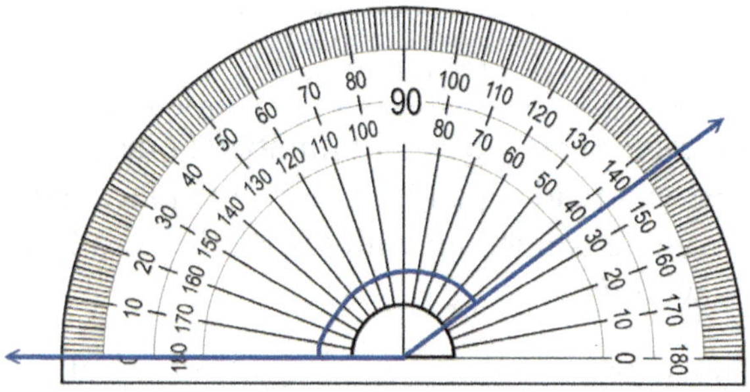

c. __________°

____

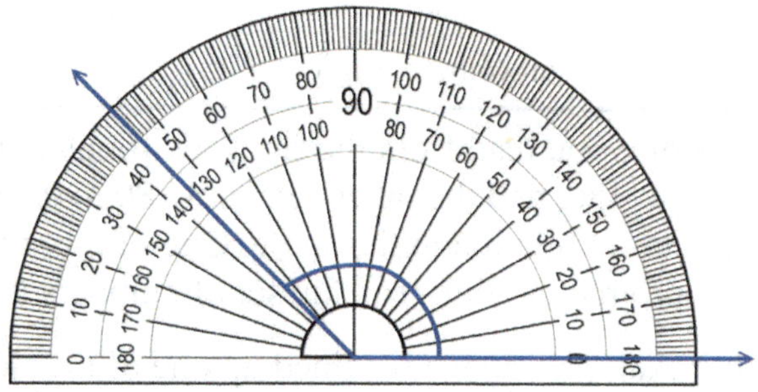

d. __________°

____

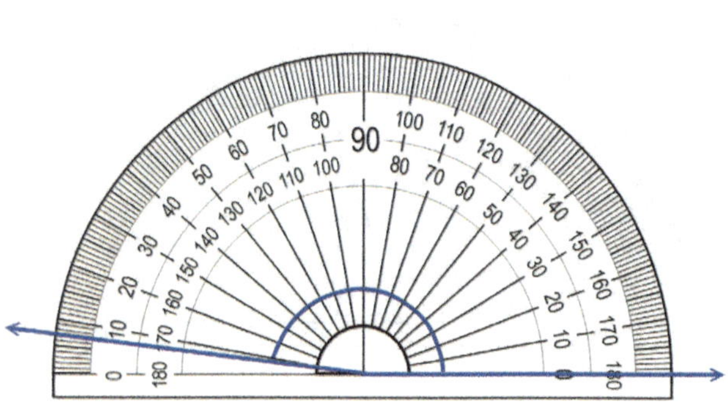

e. __________°

____

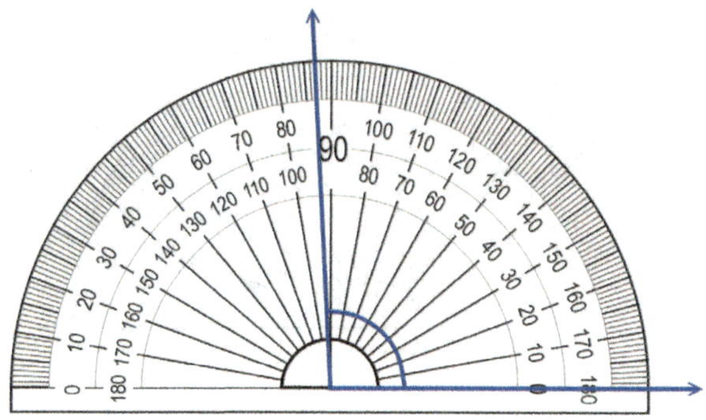

f. __________°

____

3. Tosha measured an acute angle and got 146°. The teacher pointed out that she had read the wrong set of numbers on the protractor. What is the correct measure for her angle?

4. Measure the following angles using your own protractor. If you need to, make the sides of the angles longer with a ruler.

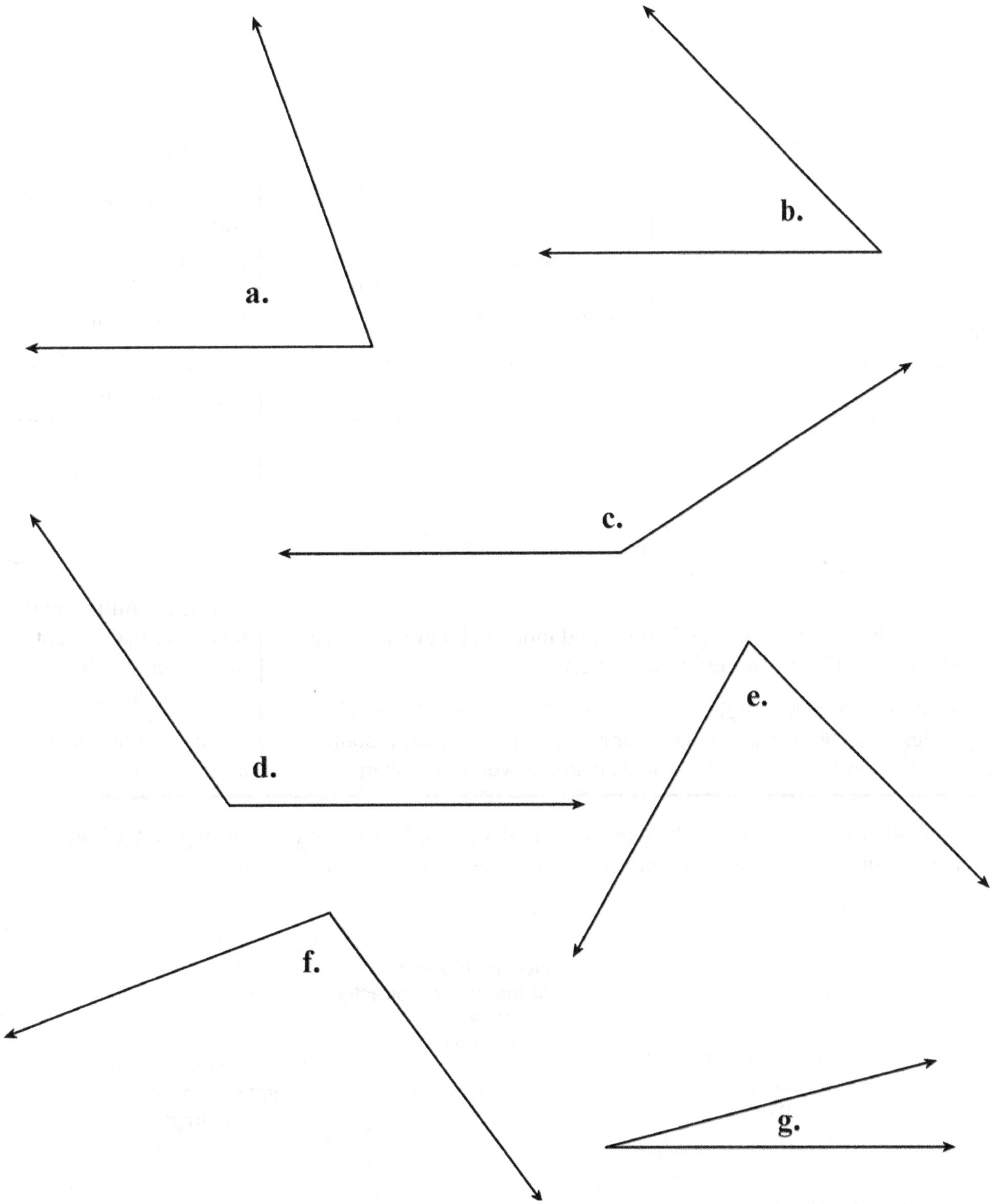

Angle terminology

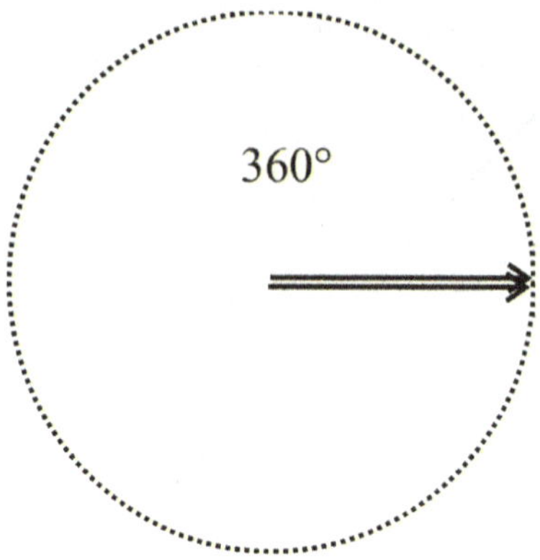

Here the pencil has drawn a full circle. This angle measures 360 degrees (360°).

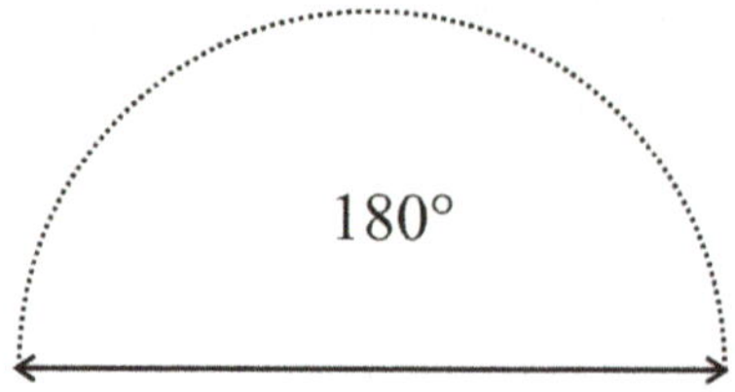

This angle is half of the full circle, so it measures 180°. It is called **the straight angle**.

Your two pencils (rays) are lying down flat or *straight* on the floor.

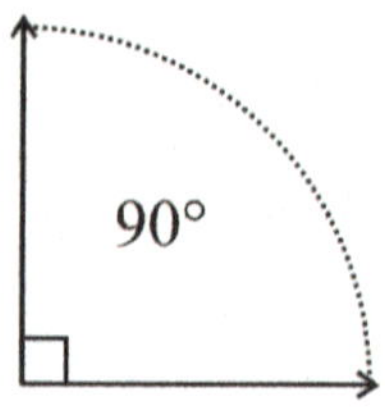

This is one-fourth of the full circle, so it is 90°. It is called **the right angle.** Table and book corners are right angles.

We often mark a right angle with a little corner.

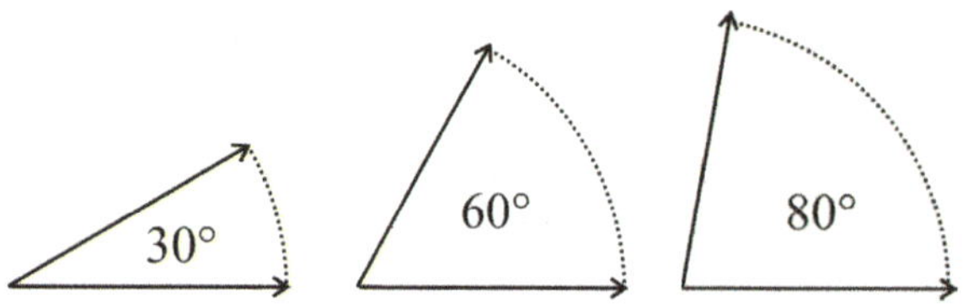

In each of these pictures the angle is opened more and more and keeps getting bigger. The arc of the circle is larger.

These angles are **acute angles**, which means they are less than a right angle (less than 90°). Think of acute angles as *sharp* angles. If someone stabbed you with the vertex of an acute angle, it would feel sharp.

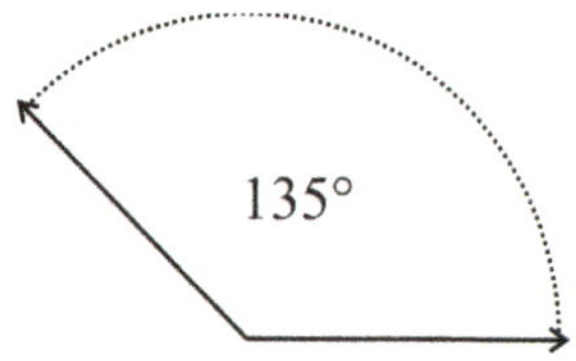

This is an **obtuse angle**: it is more than a right angle, yet less than a straight angle.

Think of obtuse angles as *dull* angles.

Here is another way of thinking about angles. Think of a SUN rising in the morning on the horizon, gradually getting higher, and traveling through the sky along an arc of a circle.

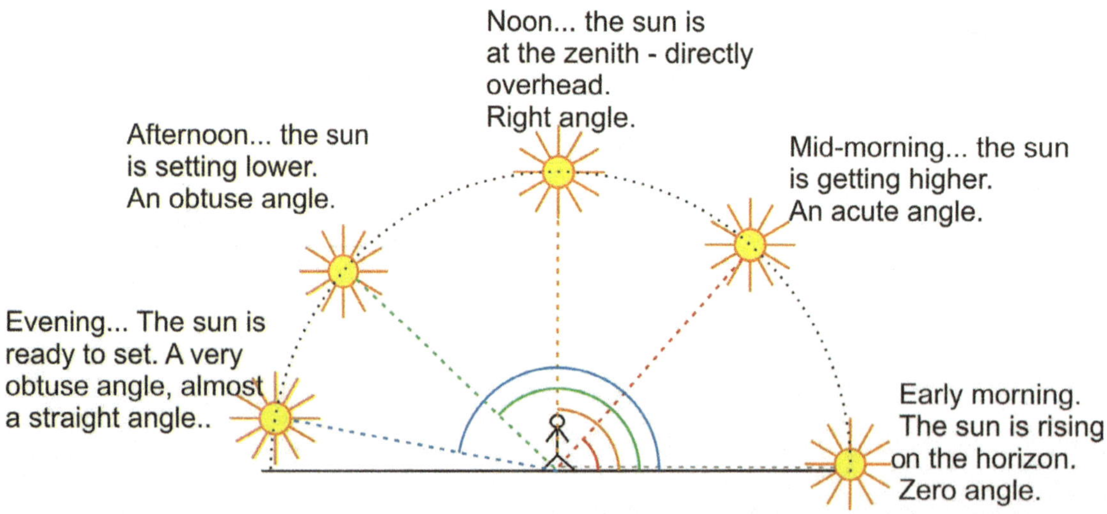

5. Label the angles as acute, right, obtuse, or straight. To help you, make these angles with two pencils, checking how much you need to open the pencils to form the angle.

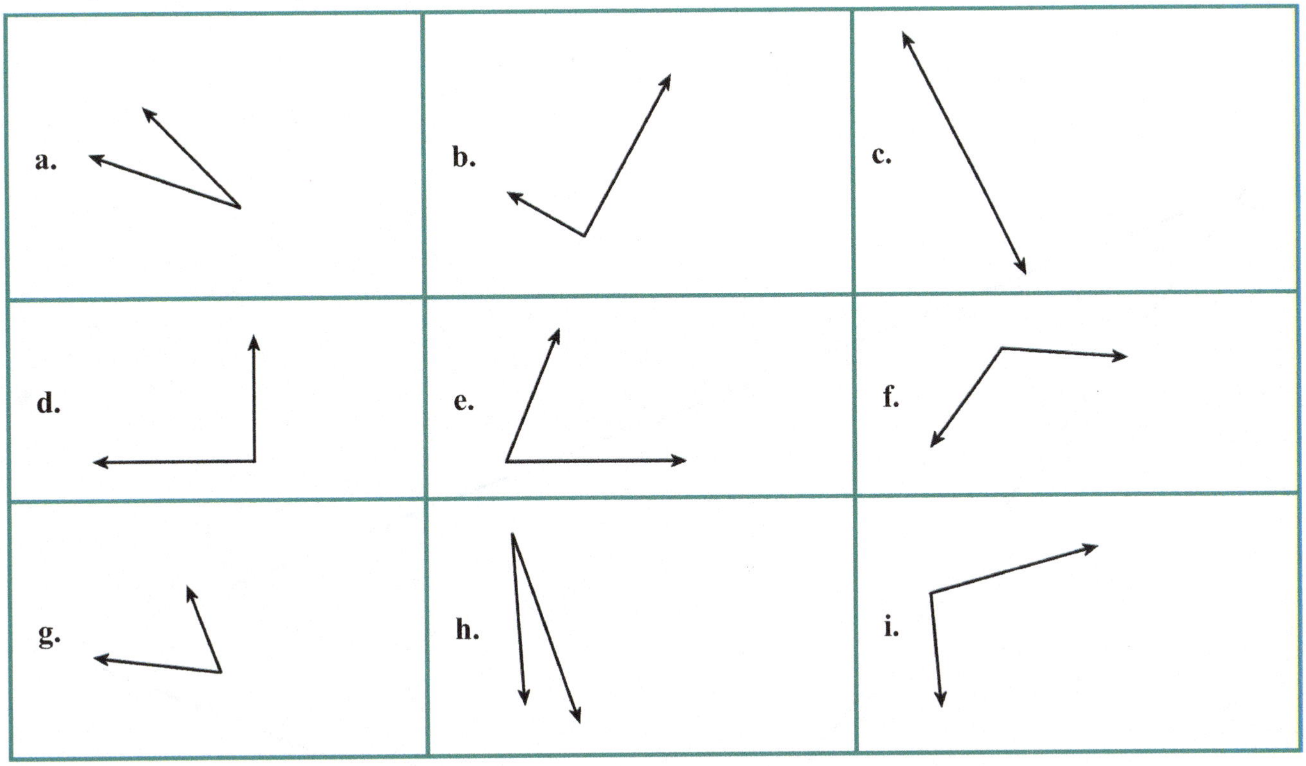

6. **a.** Sketch three different acute angles.

b. Sketch three different obtuse angles.

c. Sketch a right angle and a straight angle.

7. Measure the three angles in this triangle. Then tell if each angle is acute, right, or obtuse.

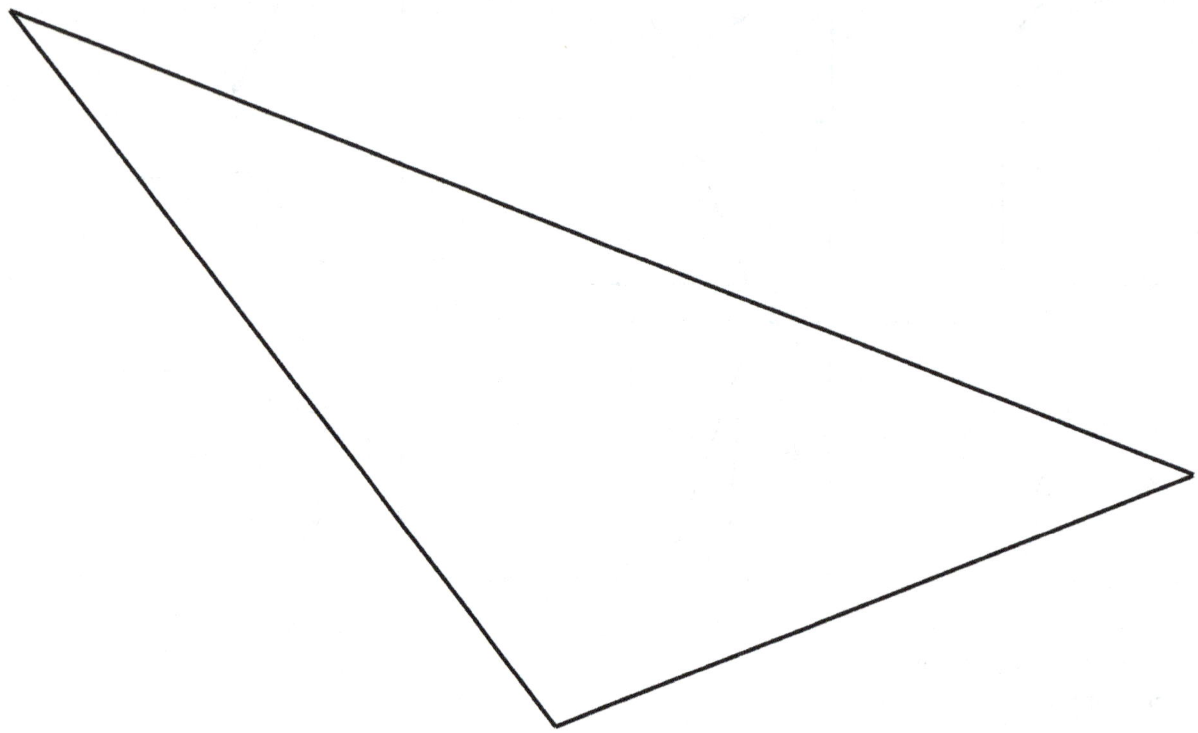

8. (Optional) Make a *geometry notebook* where you write each new term and draw a picture or pictures that illustrate the term. Use colors and write neatly. It is like your personal geometry dictionary. You can also do any drawing problems from the lessons in it. Drawing and writing them yourself, instead of just reading, can help you remember the terms better!

New Terms
• *an acute angle* • *a right angle* • *an obtuse angle* • *a straight angle*

Drawing Angles

To draw an angle of 50°, first draw a line segment. It will be one side of the angle.

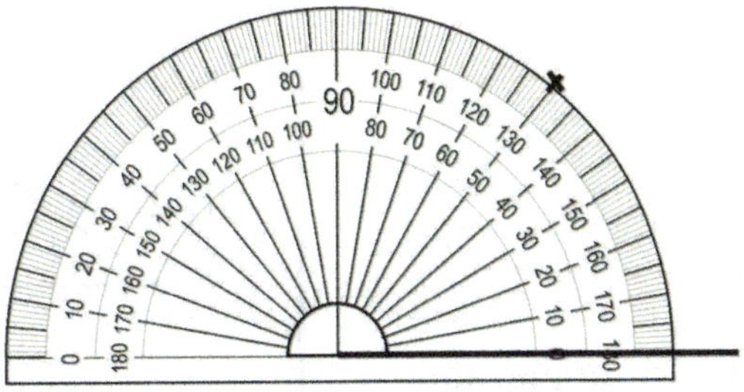

Then place the protractor so that its zero line matches with your line segment and that the vertex is in place. Then put a little mark at the 50° spot.

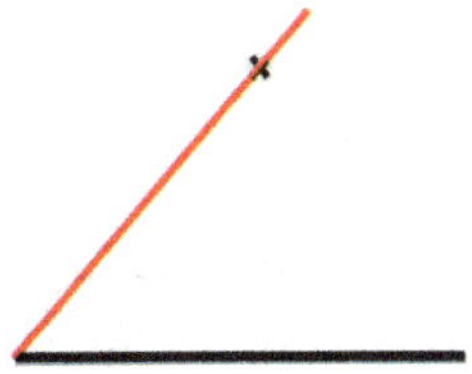

Take the protractor off and draw a line through your mark.

1. Draw angles for these measurements. You may also use blank sheet of paper.

   a. 35° **b.** 76° **c.** 137° **d.** 162°

How to draw a triangle with two given angle measurements

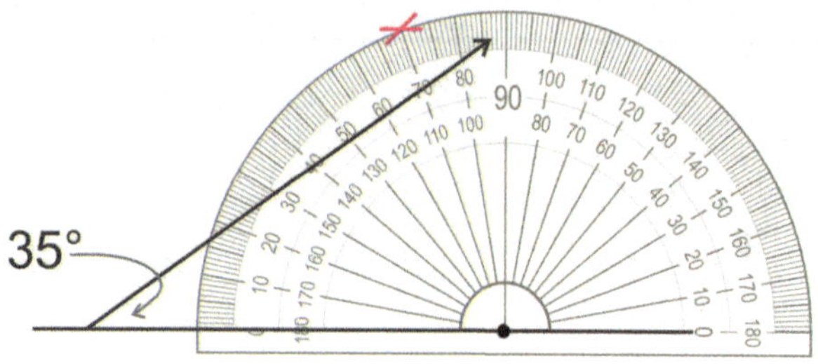

Let's say you have already drawn a 35° angle, and the second angle is supposed to be 70°. The image shows you how to place your protractor so you can measure and mark the 70° angle.

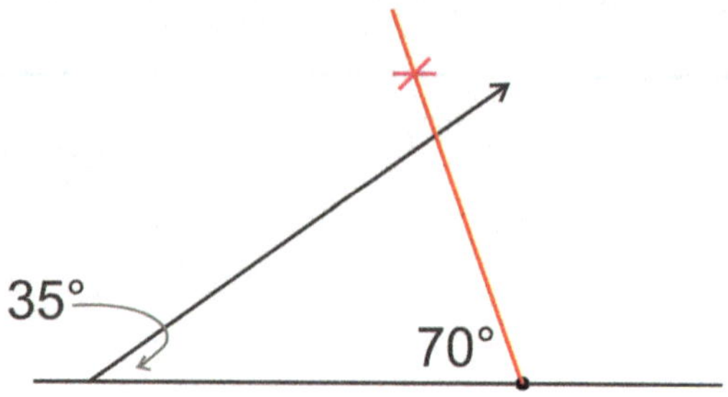

Then remove the protractor and draw the third side of the triangle.

2. Draw a triangle with 85° and 25° angles. The 85° angle is already drawn for you.

   Hint: Next, draw a point anywhere on one side of the 85° angle to be the second vertex of the triangle. Draw the 25° angle using that point as a vertex.

3. Draw a triangle with 30° and 60° angles on a blank sheet of paper.

4. If you need more practice drawing angles, draw some more angles in your notebook or on blank paper. If your teacher so directs, you can also practice drawing more triangles with given angle measurements.

5. *(Optional)* This is an angle guessing game for 2-6 people. At his turn, each player draws an angle. Everyone else will guess how many degrees it is. Then the angle is measured. The person who guessed the closest gets a point. The winner is the first person to reach 10 points.

Estimating Angles

30° and 60° angles

The angle between two neighboring clock hours is 30°.

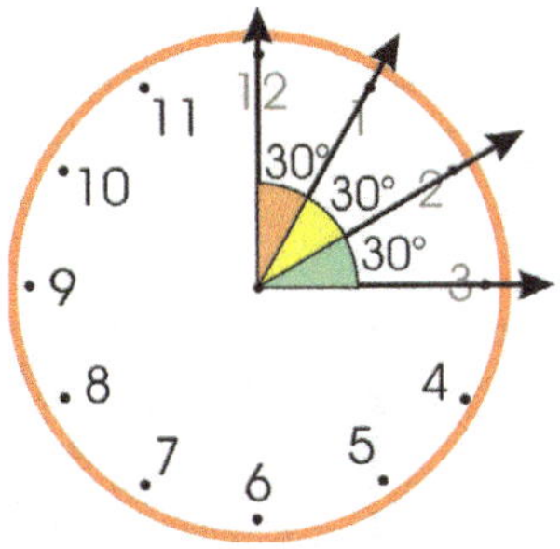

45° angle

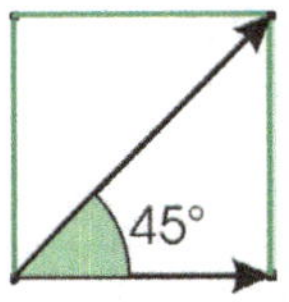

A diagonal of a square forms a 45° angle with the side of the square.

120° angle

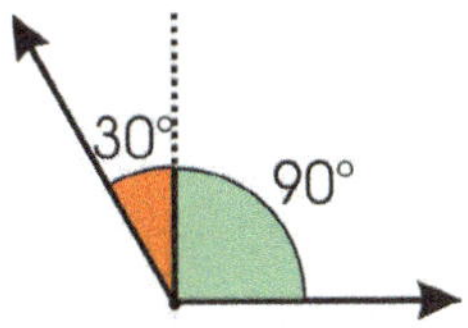

Add 30° to a right angle.

150° angle

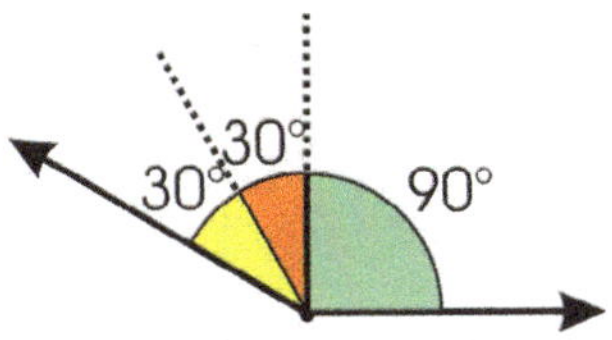

Add yet another 30° angle.

135° angle

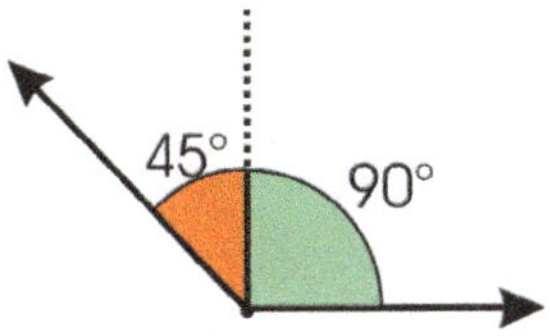

Add a 45° angle to a right angle.

1. **a.** It is two o'clock. What angle is formed by the minute hand and the hour hand?

   b. It is five o'clock. What angle is formed by the minute hand and the hour hand?

2. Estimate the angle measures in degrees.

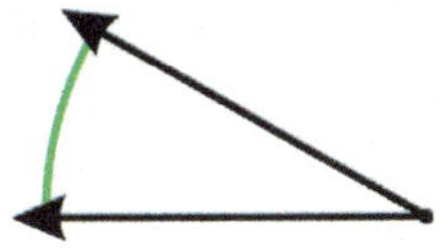

a. ____________

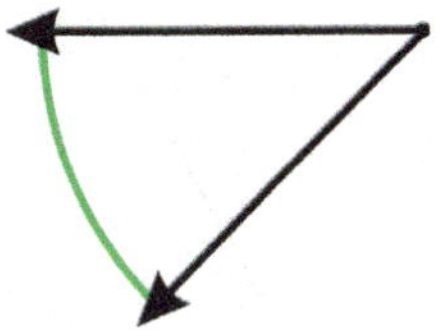

b. ____________

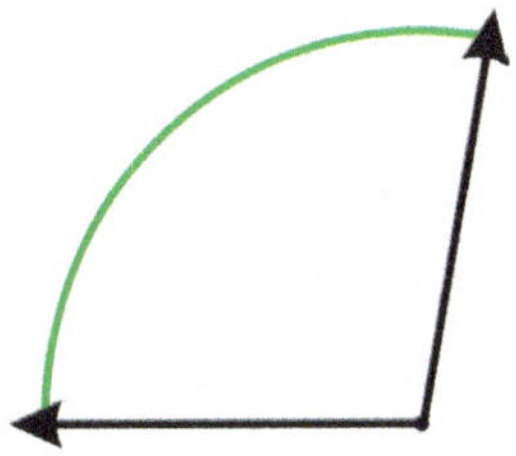

c. ____________

d. ____________

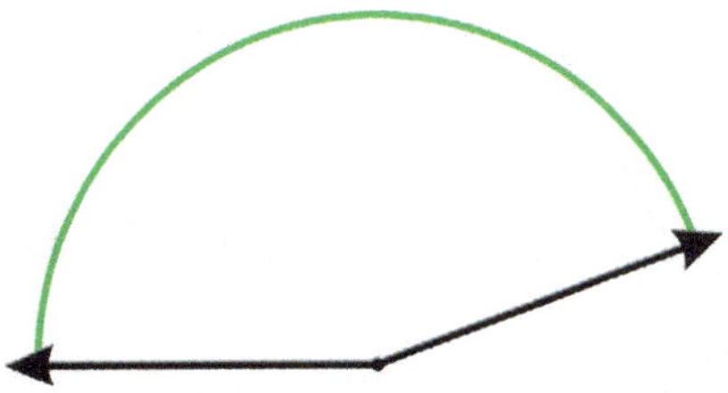

e. ____________

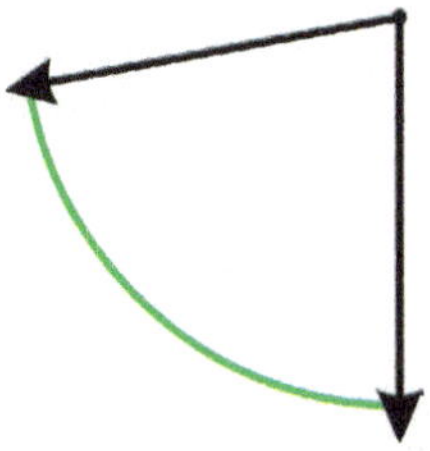

f. ____________

3. Estimate the angle measures in degrees. Note that the last two angles are bigger than 180°—they are **reflex angles**.

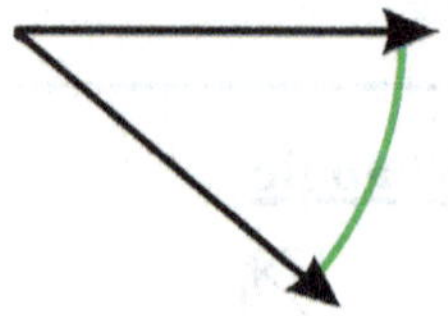 **a.** ______________	 **b.** ______________	 **c.** ______________
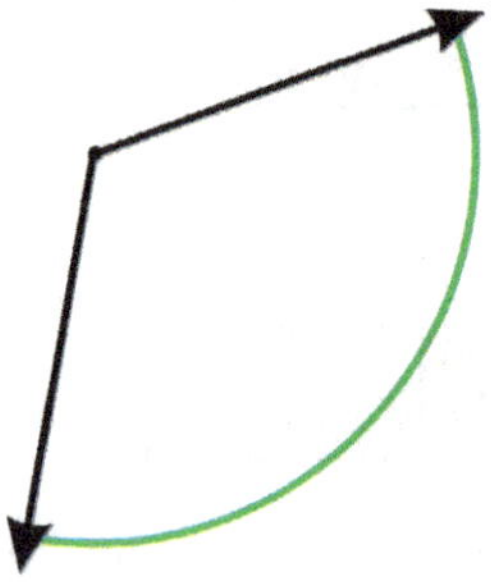 **d.** ______________	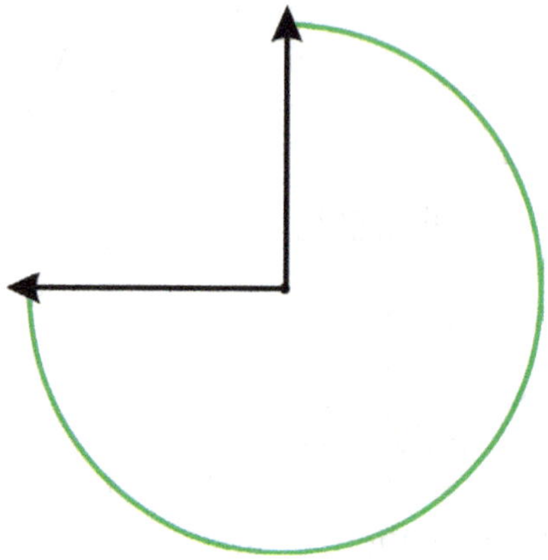 **e.** ______________	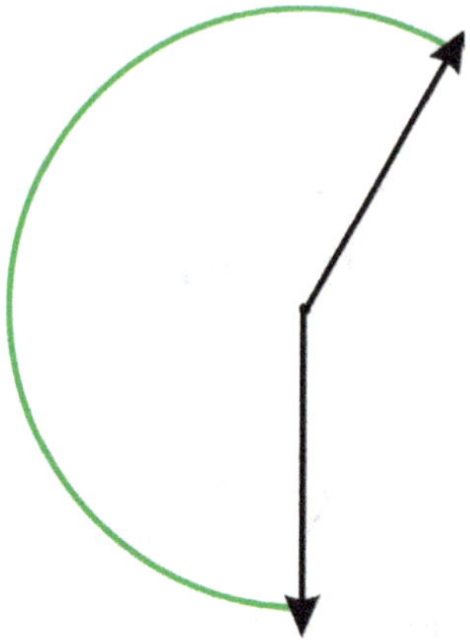 **f.** ______________

4. Estimate the measures of all the angles in this triangle. Then measure with a protractor to check.

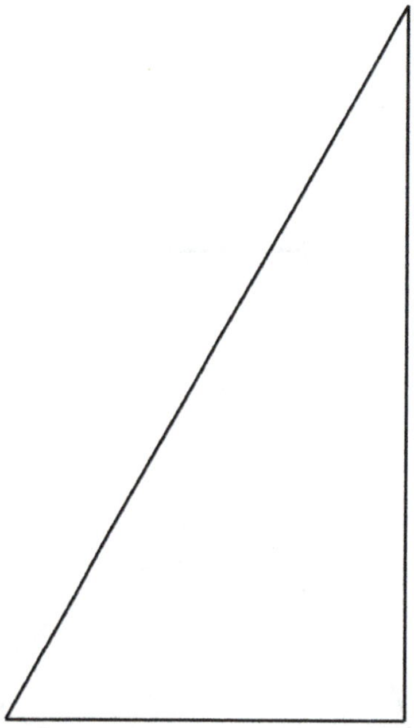

5. Draw these angles without using a protractor. In other words, estimate. After you have drawn the angle, measure it with a protractor to see how close you came to the correct angle.

a. 45°	**b.** 60°
c. 30°	**d.** 120°

Turning in an angle (*This section is optional.*)

Look at the pictures. They show turning at a certain angle.

The two sides of the turning angle are the direction you *would have continued*, and the direction you actually go to.

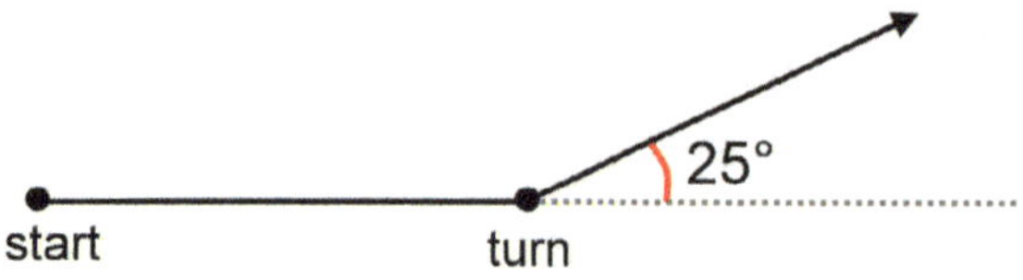

Turning 25° to the left.

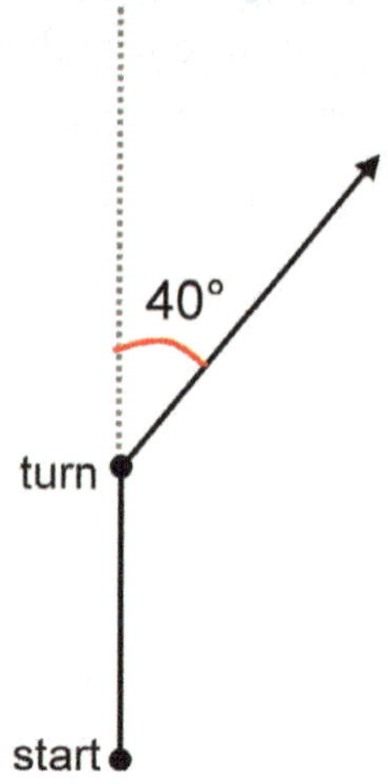

Turning 40° to the right.

6. First estimate the turning angle and find which toy the child would get. Then check with a protractor.

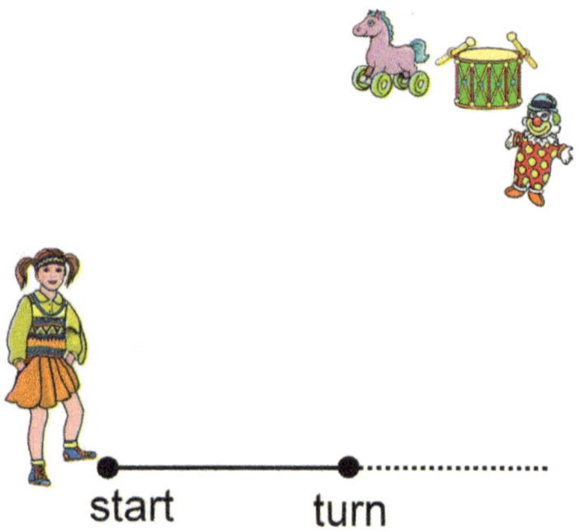

a. Jeanine turns 70° to her left.

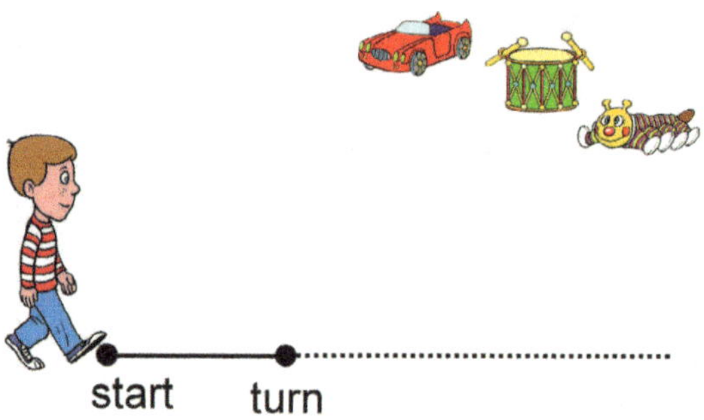

b. Jack turns 30° to his left.

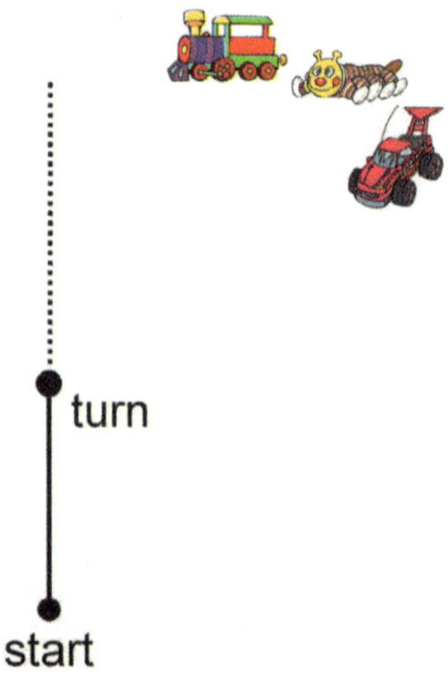

c. Turn 30° to the right.

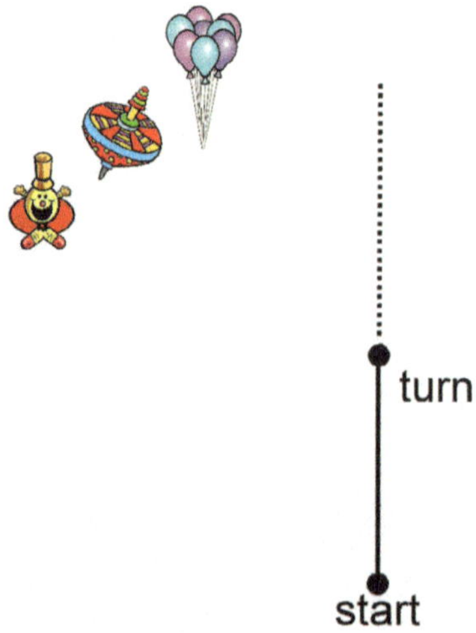

d. Turn 50° to the left

7. Sketch the route of a person who travels and then turns in a 45° angle to his right.

8. Sketch the route of a person who travels and then turns in a 60° angle to his left.

Puzzle Corner

Follow the directions, and "gather" all the letters you pass through. What message will they make?

Start. Move five blocks to the north (up). Turn 30° to the right and go until you hit a letter. Turn due east. Go three blocks. Turn 45° to your right and go until you hit a letter. Turn due west. Go two blocks. Turn 30° to your left and continue till you hit a letter. At the letter, turn due south and continue four blocks. Turn 90° to your left and go two blocks. Turn 45° to your left and continue until the letter "R". Stop.

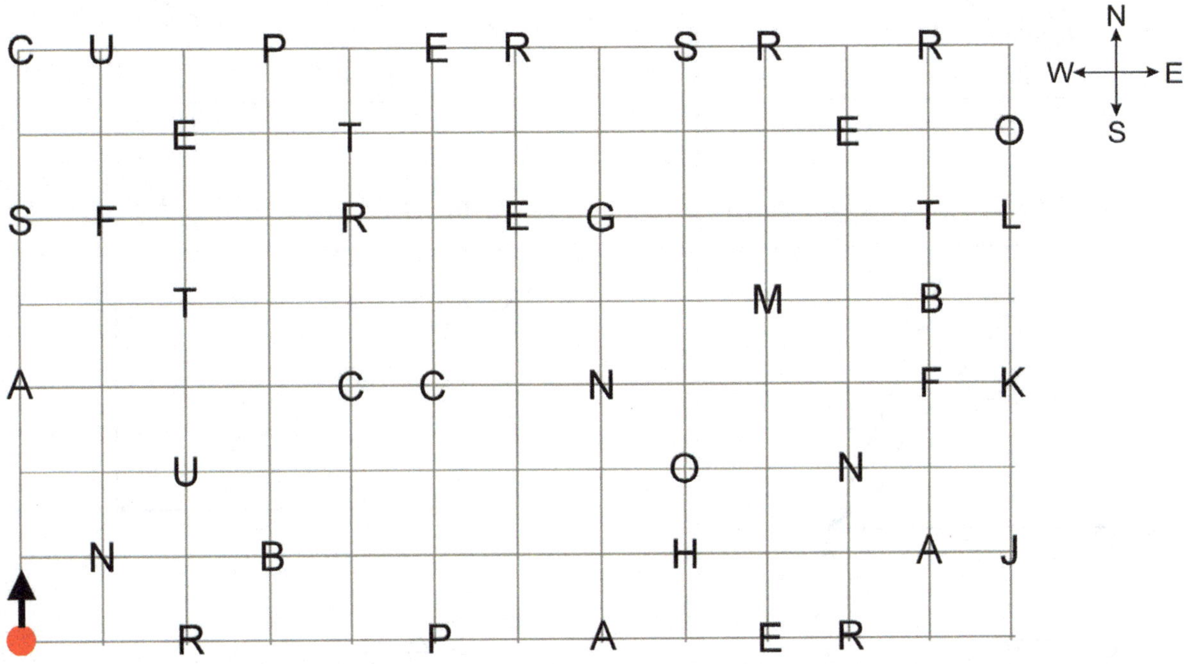

Angle Problems

1. Angle ABC is a right angle.

Think: can you *calculate* ∠ABD if you first measure∠DBC?

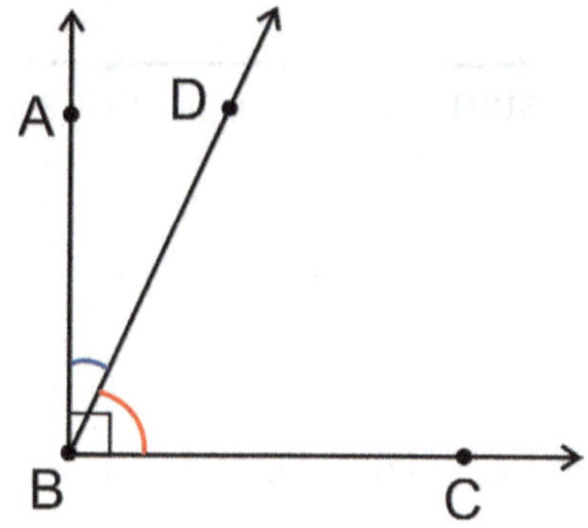

We can ADD angle measures when the angles share a vertex and they do not overlap.

Therefore,

∠BAD = 20° + 44° = 64°

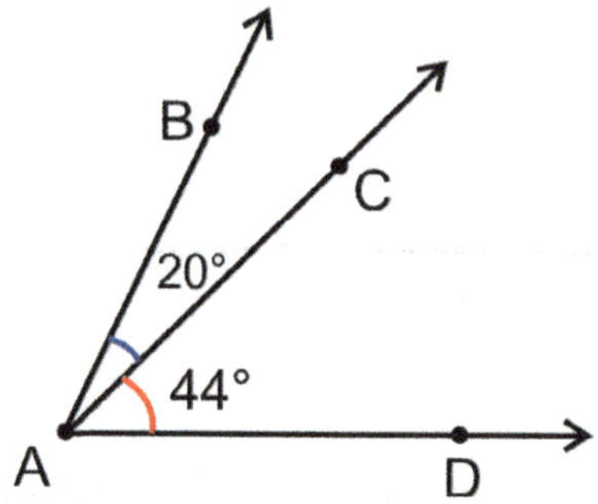

2. Figure out the unknown angle. Do *not* measure it. Use logical thinking!

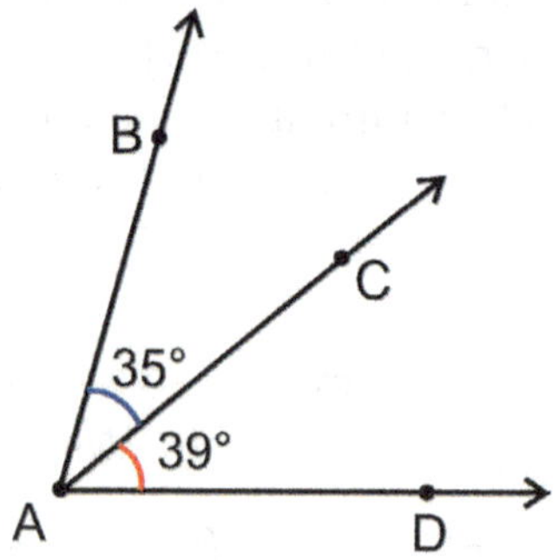

a. ∠BAD = ________ °

Note:
∠BAD = 81°

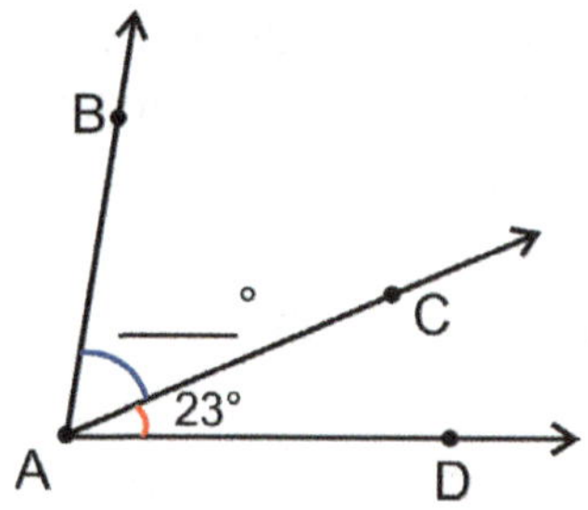

b. ∠BAC = ________ °

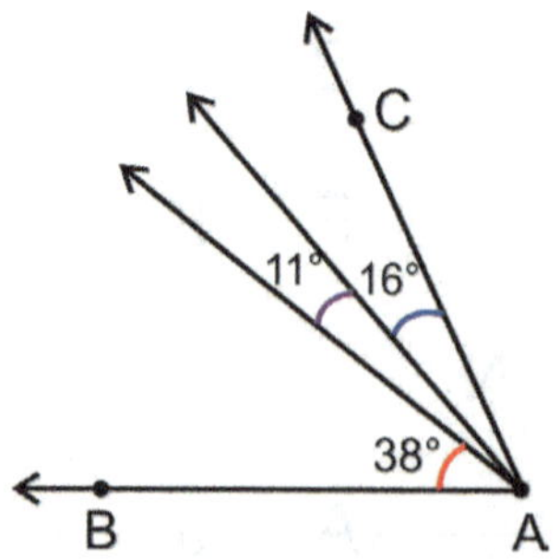

c. ∠BAC = ________ °

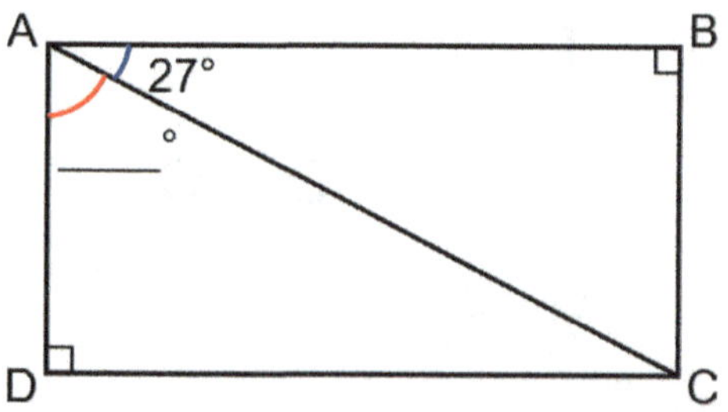

d. ∠CAD = ________ °

3. Figure out the unknown angle. Do *not* measure it.

Note:
∠BAD = 132°

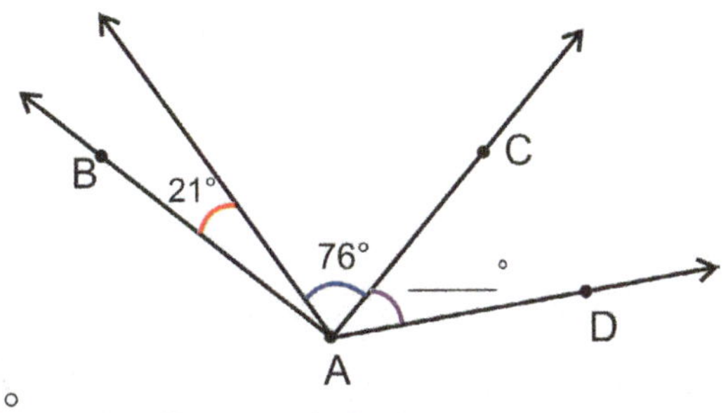

a. ∠CAD = ________°

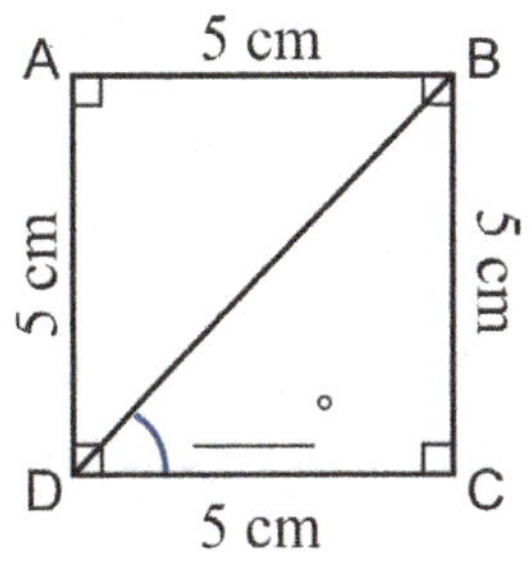

b. ∠BDC = ________°

4. Write an addition for the angle measures. Use the unknown x for one angle measure. Then find what the unknown stands for. The first one is done for you.

Note:
∠BAD = 72°

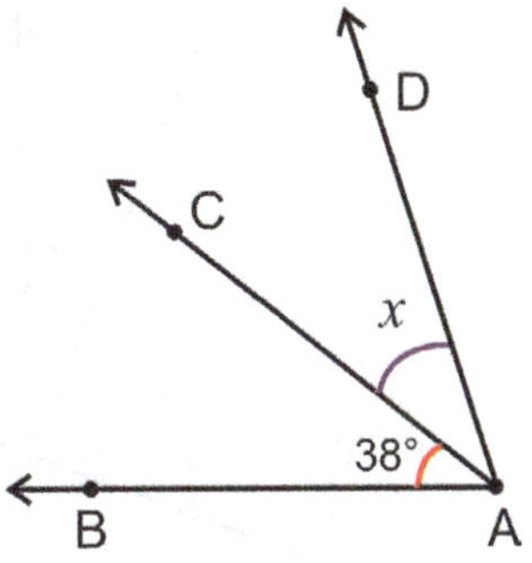

a. $38° + x = 72°$

$x = 34°$

Note:
∠BAD = 90°

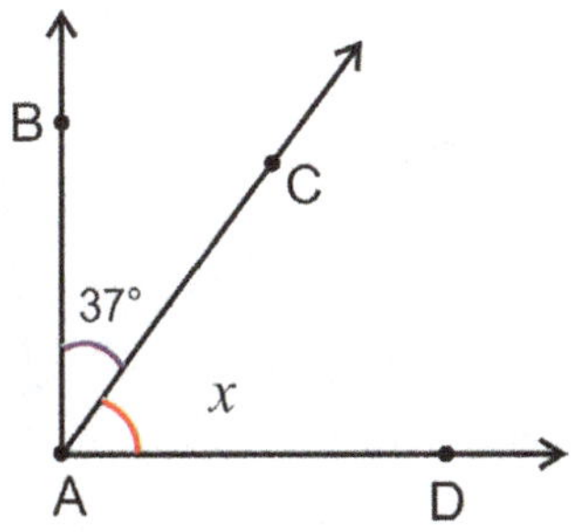

b. ______________________________

____

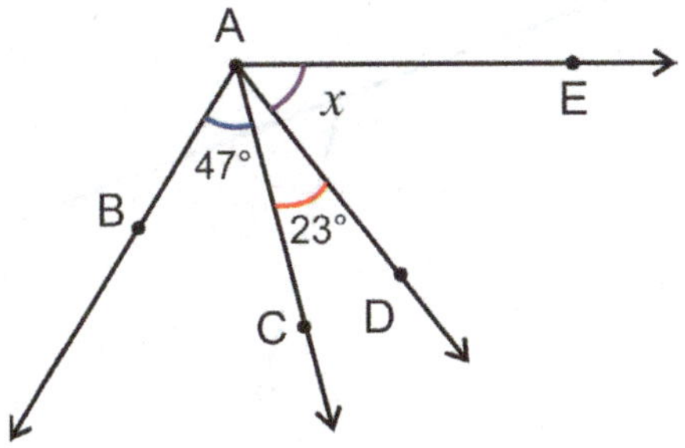

Note: **∠BAE = 122°**

c. ______________________________

____

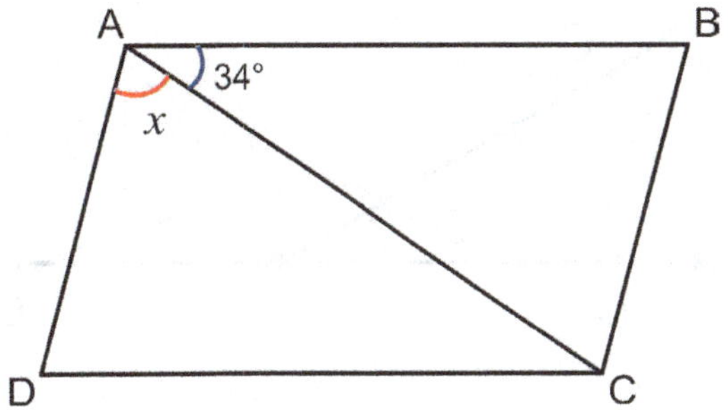

Note: **∠BAD = 105°**

d. ______________________________

____

In this picture, we use a double arc to mark ∠CAD. It is used for clarity and simply marks an angle, just like a single arc does.

Angle BAD is a straight angle,

so it measures _________°.

Now think. Can you calculate ∠BAC if you first measure ∠CAD?

Yes, you can!

You can simply subtract the measure of ∠CAD from 180°, and you will get the measure of ∠BAC.

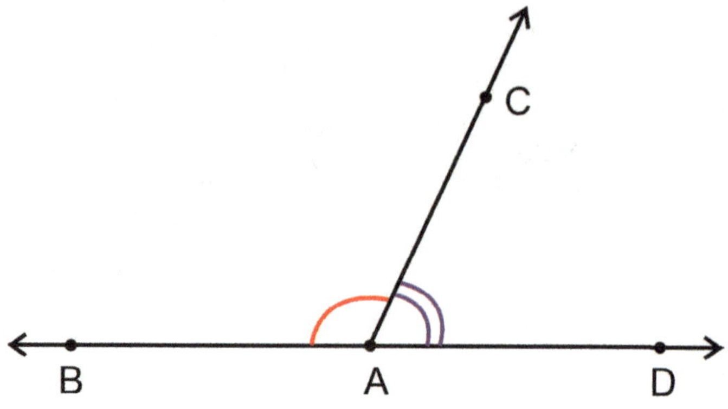

5. Figure out the unknown angle.

a.

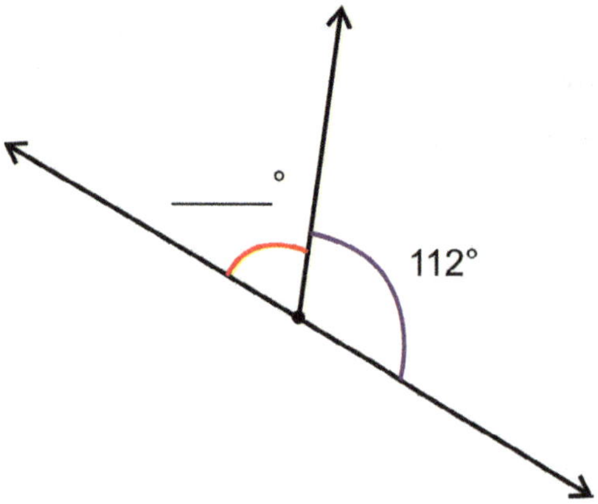

b.

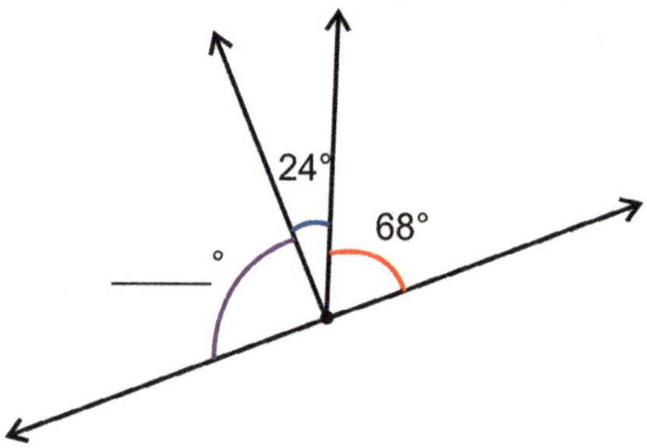

6. Write an addition for the angle measures. Use the unknown x for one angle measure. Then find the value of x.

a.

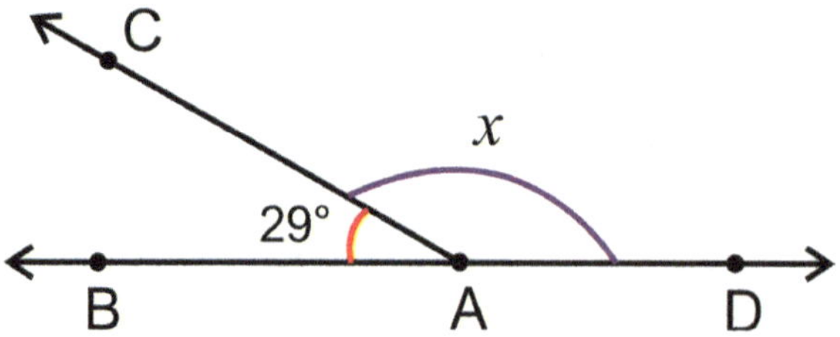

____

____

____

b.

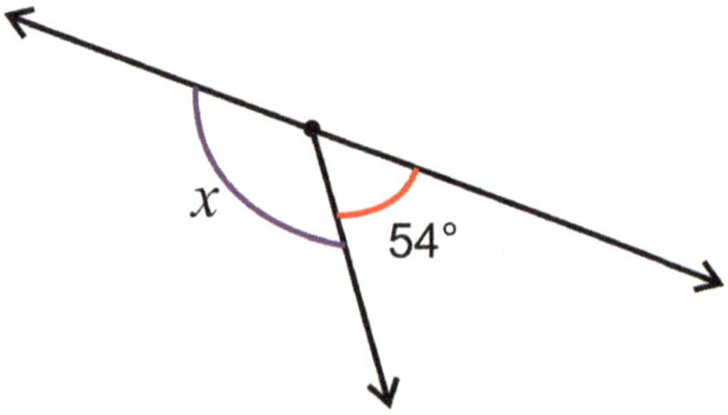

____

____

____

There are three angles in this picture that share the same vertex. Notice that these three make a complete circle!

Can you guess what their angle measure is in total?

When many angles share a vertex forming a complete circle, the sum of their measures is 360°.

So, the sum of these angle measures is 360°. In symbols, $x + y + z = 360°$.

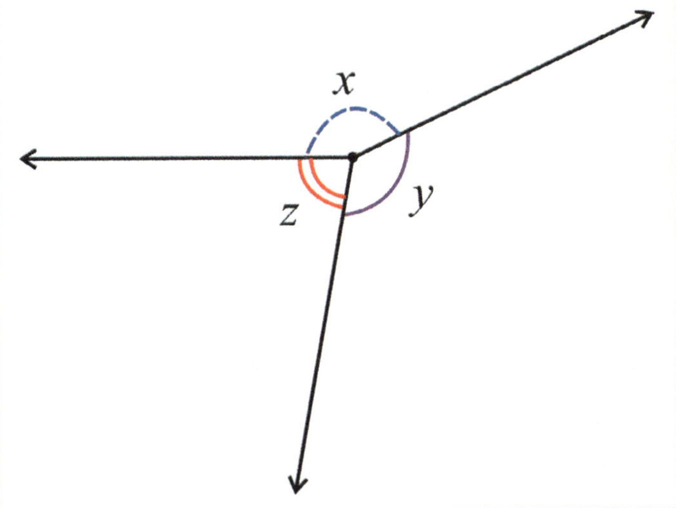

7. Write an addition for the angle measures. Use the unknown x for one angle measure. Then solve the equation you get.

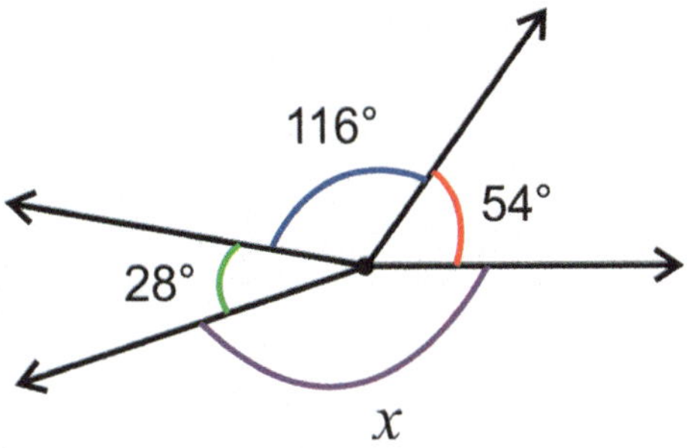

a.

____

____

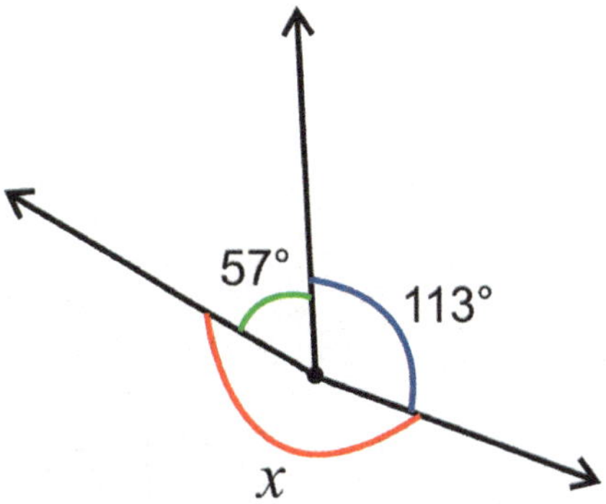

b.

____

____

8. What is the angle sum of these three angles? One angle is a straight angle. Another measures half of the first. The last angle measures 1/3 of the second angle.

9. Challenge*:* Find the measure of the *marked* angle.
   Note: it is greater than 180°!
   We call such angles **reflex angles**.

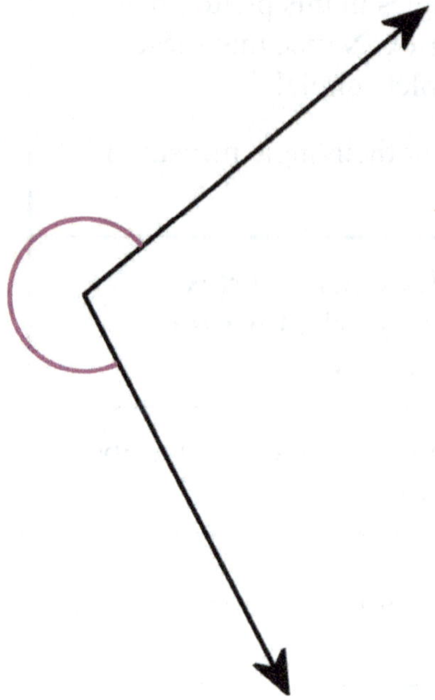

Puzzle Corner

a. A circle is divided into eighths. Find the measure of the marked angle.

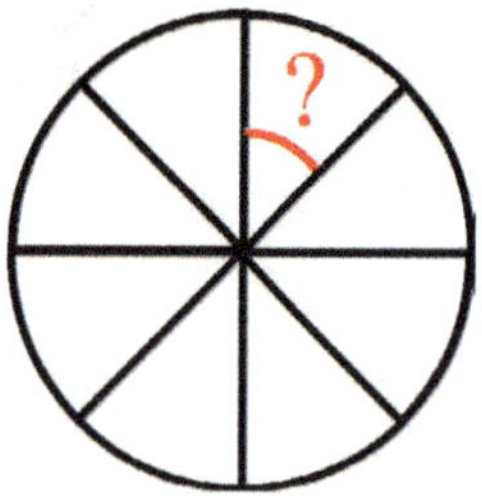

b. Find the measure of the marked angle.

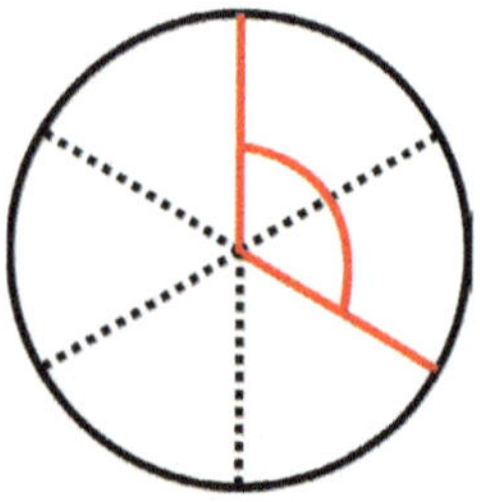

Parallel and Perpendicular Lines

Two lines or line segments can either **intersect** (cross each other) or be **parallel.**

Think of two parallel lines as never meeting each other, no matter how much you would continue them in both directions.

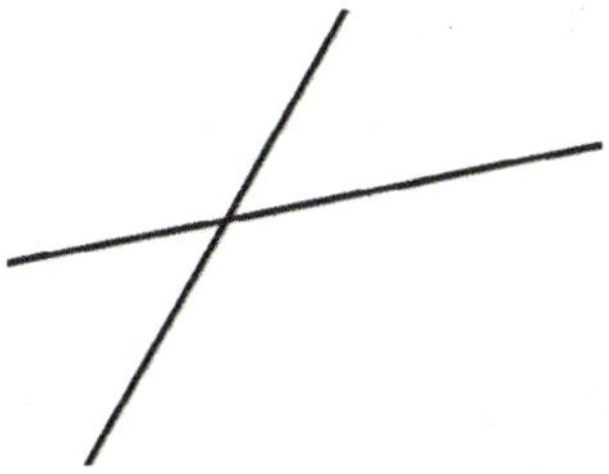

These lines intersect.

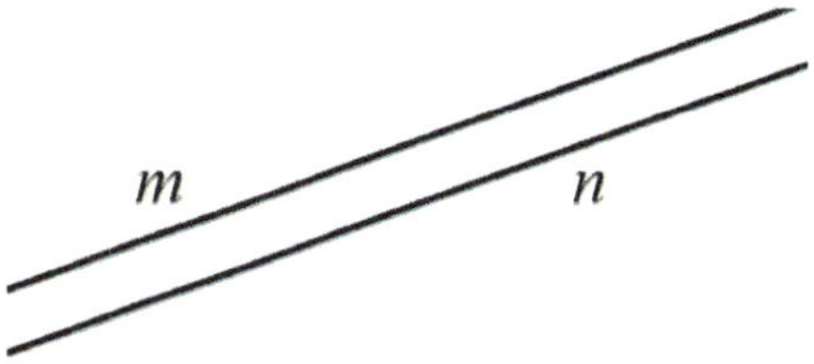

Lines m and n are parallel.
We write $m \parallel n$.

Two lines or line segments are **perpendicular** if they **form a right angle**.

We use the symbol ⊥ to mean "is perpendicular to".

In an image, we can mark a right angle with a little corner like this: ⌉.

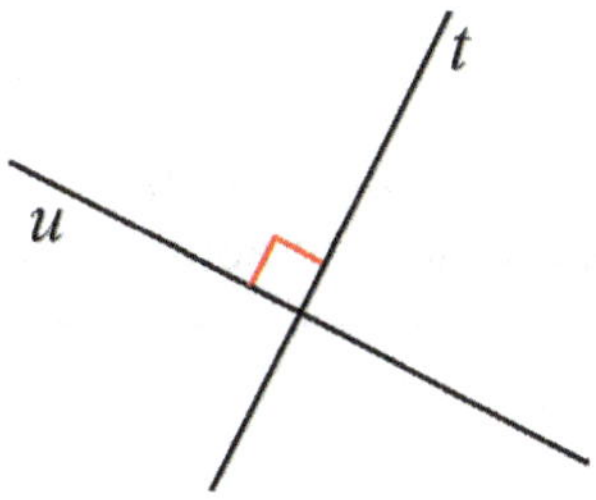

How many right angles do the lines u and t form?

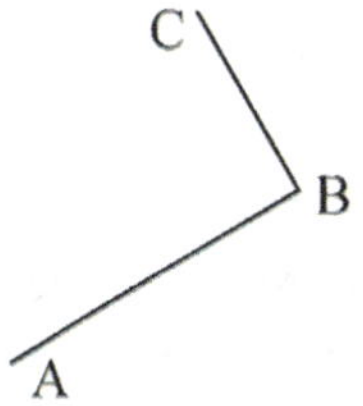

The line segments $\overline{AB}$ and $\overline{BC}$ are perpendicular.

We write $\overline{AB} \perp \overline{BC}$.

1. Do these lines intersect or are they parallel?
   Continue the lines with your ruler.

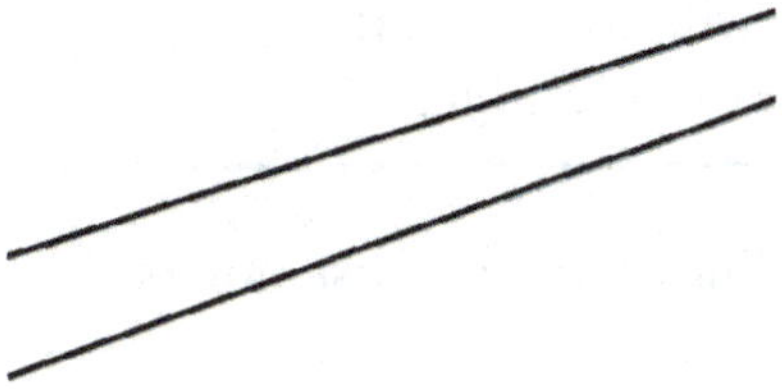

2. Which line segments in these figures are parallel? Which are perpendicular?

a. Line segments AB and BC

are ____.

Line segments AD and BC

are ____.

b. Line segments EF and GH

are ____.

Line segments EH and FG

are ____.

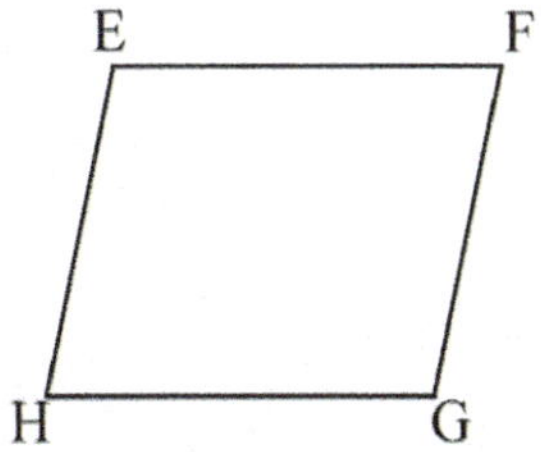

How to draw a line that is perpendicular to a given line.

First, draw a point on the given line. Next you will need a protractor or a triangular ruler.

1) Using a protractor:

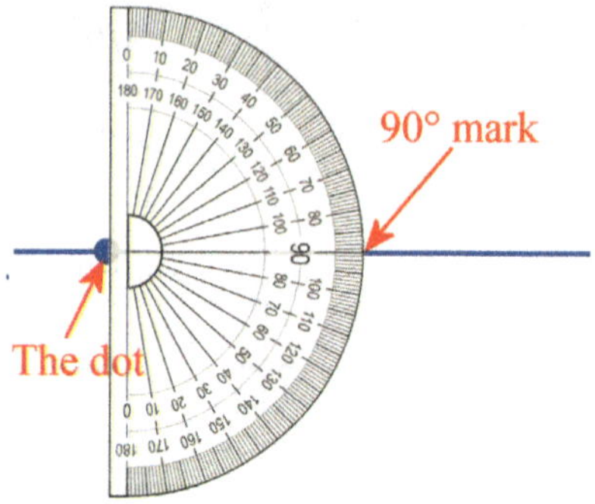

Align the dot and the straight side of your protractor. Also align your existing line and the 90° mark on the protractor.

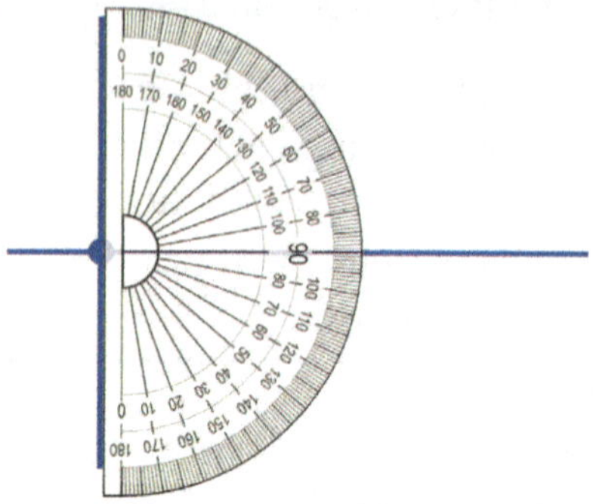

Draw the line.

2) Using a triangular ruler:

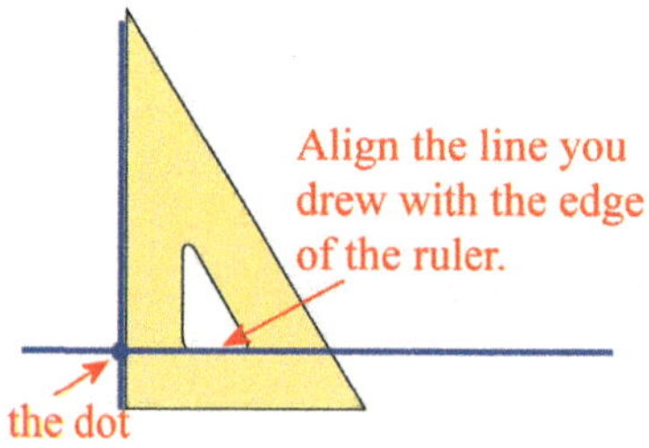

Align the *inside* edge with the given line.
Align the dot with the outside edge.

Draw the line.

3. Draw perpendicular lines through these points.

4. Draw a line that is perpendicular to the given line and goes through the given point.

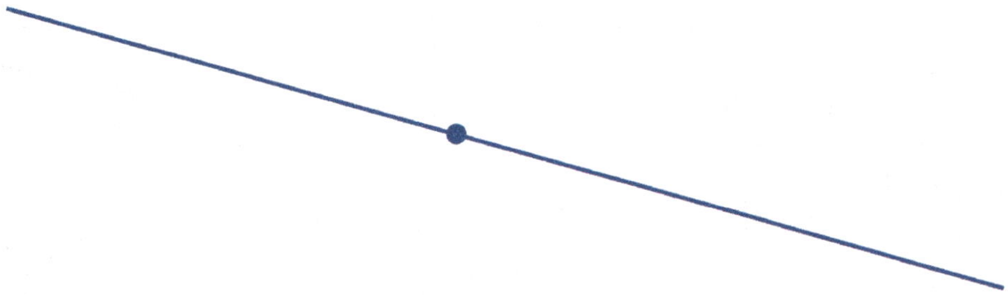

5. Complete these drawings so you get: a) a rectangle; b) a square. Use a protractor or a triangular ruler to make sure the lines you draw are perpendicular to the existing lines.
Hint: You may need to continue the existing lines.

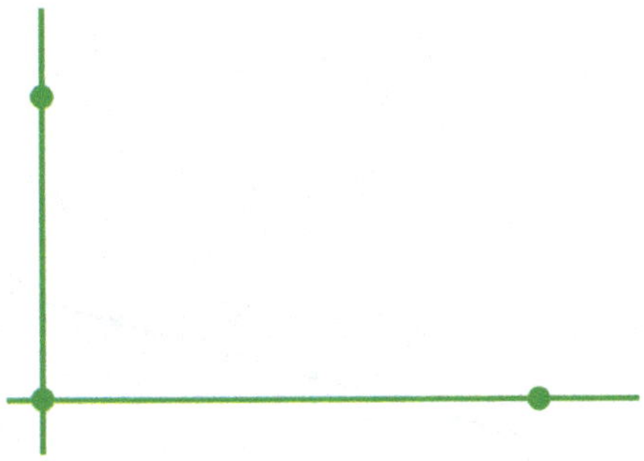

a.

b.

6. **a.** Draw here or on a blank sheet of paper any triangle that has a right angle. It is called a right triangle.
(Hint: Start by drawing two lines that are perpendicular.)

b. Find the perimeter of your triangle in centimeters and millimeters.

How to draw a line that is parallel to a given line.

Method 1: A ruler. Align the bottom side of the ruler with an existing line. Then draw a line above the top of the ruler.

You can carefully slide the ruler up or down if you need the parallel line to be further or closer to the existing line. That means the drawing may not be totally accurate, though.

Method 2: A protractor.

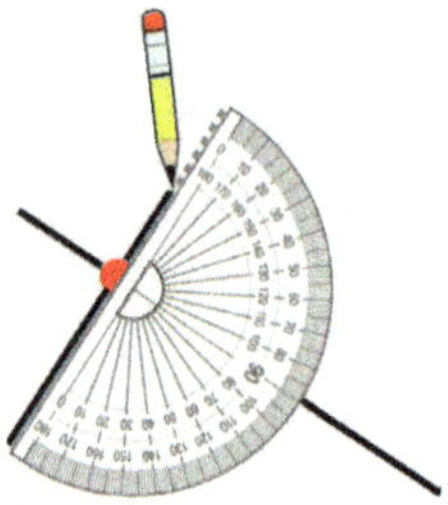

I. Draw a point on the given line. Then draw a perpendicular line through that point.

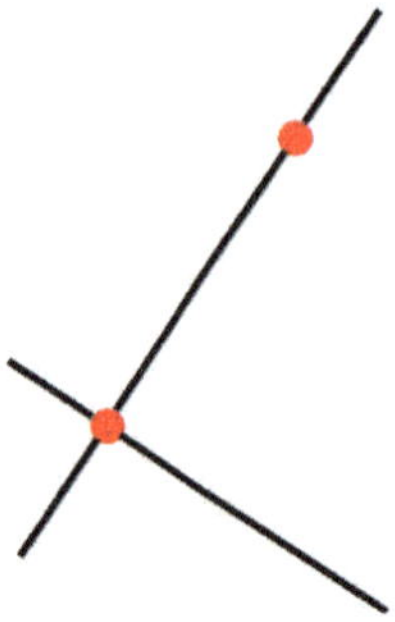

II. If the line is not long enough, continue it with a ruler. Then, draw a point on it.

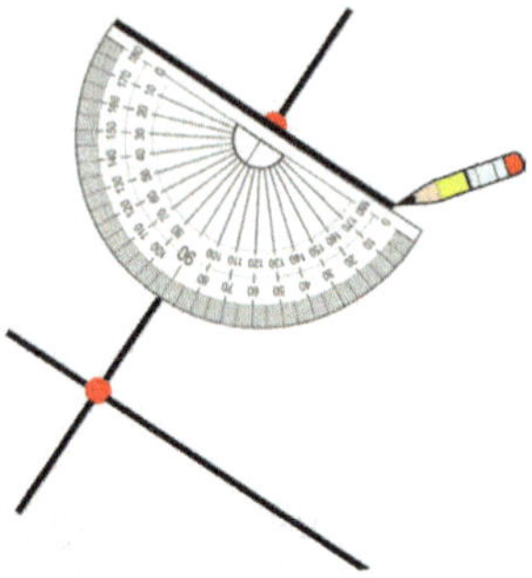

III. Draw a perpendicular line. Make that line longer using a ruler. All done!

7. Sketch several lines that are *parallel* to this line using method 1. Then draw one line that is perpendicular to them all!

8. Draw a line that is parallel to this line using method 2.

9. Determine if the line segments are parallel or perpendicular to each other.
Write ∥ for parallel and ⊥ for perpendicular.

$\overline{AF}$ ▢ $\overline{FB}$ $\overline{AF}$ ▢ $\overline{BE}$

$\overline{FB}$ ▢ $\overline{BE}$ $\overline{AB}$ ▢ $\overline{FE}$

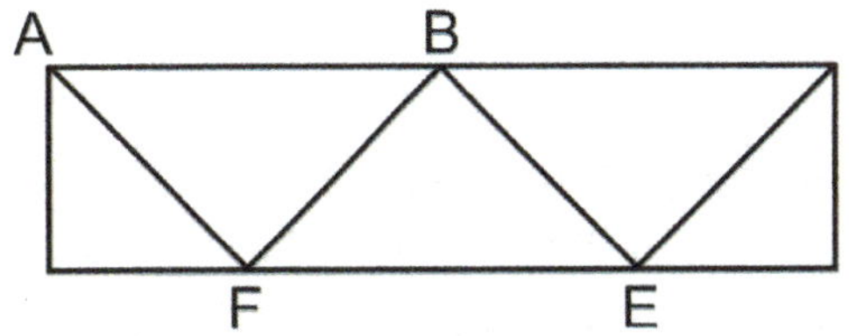

10. Find rays, lines, and line segments that are either parallel or perpendicular to each other.

a.

⊥

⊥

∥

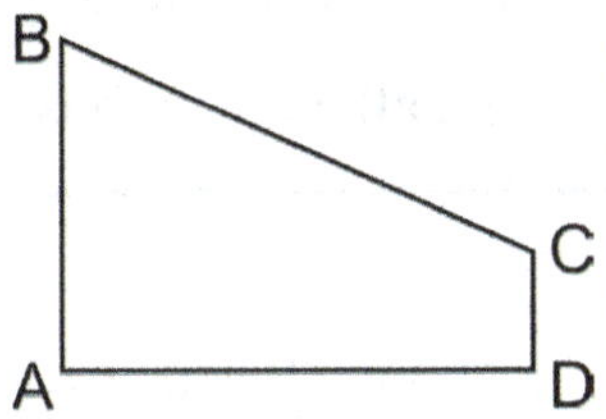

b.

⊥

⊥

⊥

∥

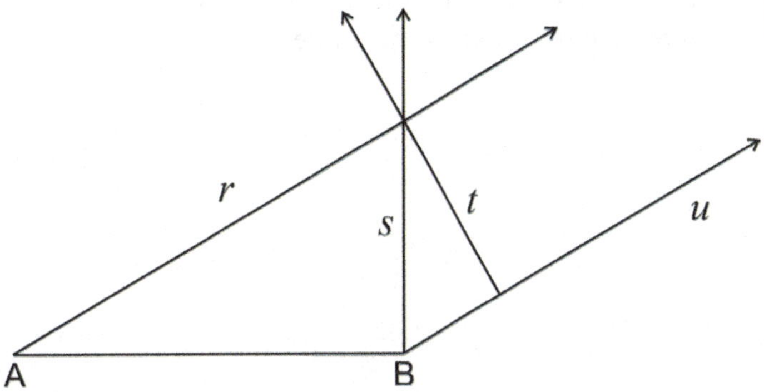

c.

⊥ ∥

⊥ ∥

⊥ ∥

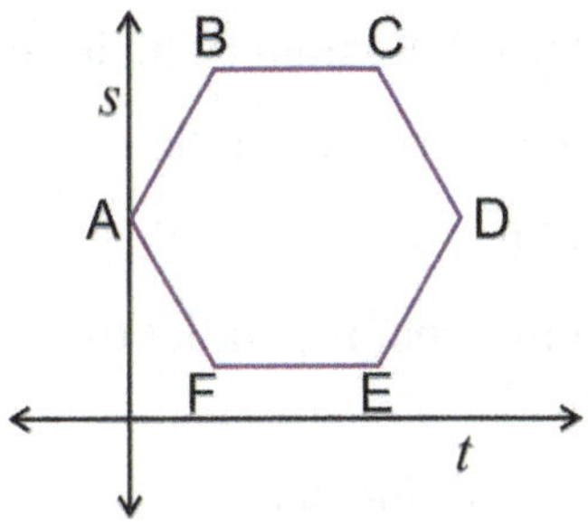

New Terms & Symbols	
• *intersecting*	• *perpendicular*
• *parallel*	• ∥ and ⊥

Parallelograms

In this picture you see two parallel lines, *a* and *b*. The third line, *t*, intersects the other two.

Draw a *fourth line* in the picture so it is parallel to the line *t*, and intersects the other two. Call it *s*.

Color the figure formed in the middle. It is a **parallelogram**, because two of its sides are parallel to each other, and so are the other two.

A parallelogram has two *pairs* of parallel sides.

1. **a.** Draw a line that is parallel to the line *s* so that you get a parallelogram.

   b. Measure all the sides of the parallelogram in millimeters. Write the side lengths in the figure, next to each side.

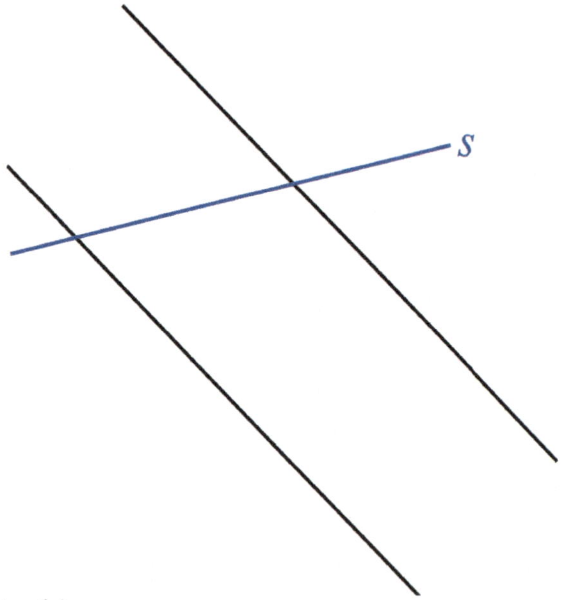

2. Draw any parallelogram (with a ruler and possibly other tools).

   The following outline can help.

   1) Draw a line.

   2) Draw a line that is parallel to that line.

   3) Draw a third line that intersects the other two lines.

   4) Draw a fourth line that is parallel to the third line.

If *all* sides of a parallelogram are congruent (of the same length), it is called a **rhombus**.

In common speech, rhombi (plural of rhombus) are also called diamond-shapes or diamonds.

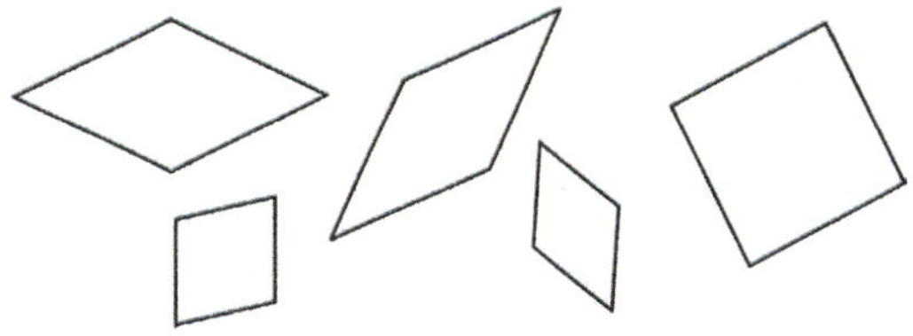

On this web page you can play with rhombi interactively:
https://www.mathsisfun.com/geometry/quadrilaterals-interactive.html

3. Draw several different rhombi in the grid. Vary the sizes and angles. You can also make a pattern with them!

If two sides of a quadrilateral are parallel, it is called a **trapezoid** (trapezium in British English).

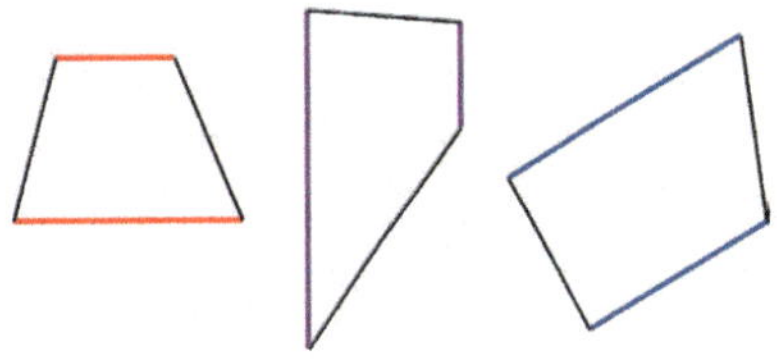

4. Draw a bunch of different trapezoids in the grid! Hint: start out by drawing two parallel lines, of different lengths.

5. Label each quadrilateral as a rhombus, a trapezoid, or a parallelogram.

a. 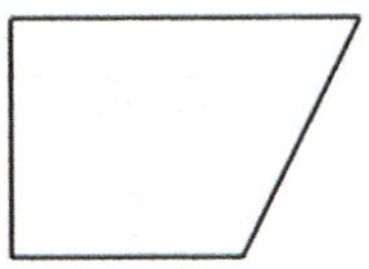	**b.** 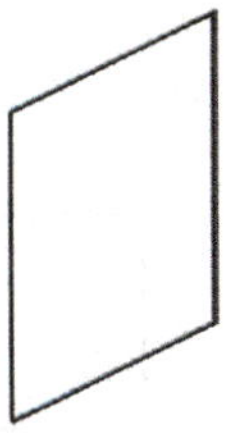	**c.** 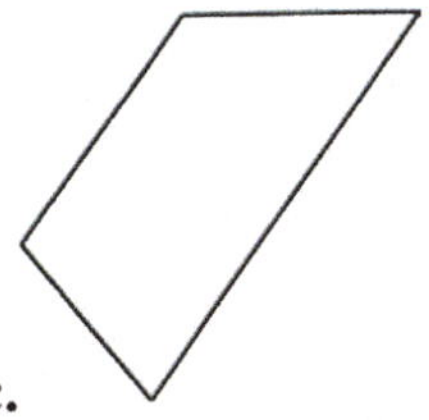	**d.** 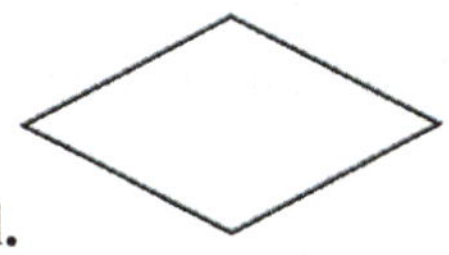

6. **a.** Find the midpoint of each side of this rectangle, accurately. *(How? Measure with a ruler.)* Carefully join the midpoints of adjacent sides with line segments (use a ruler).

b. What figure is formed in the middle?

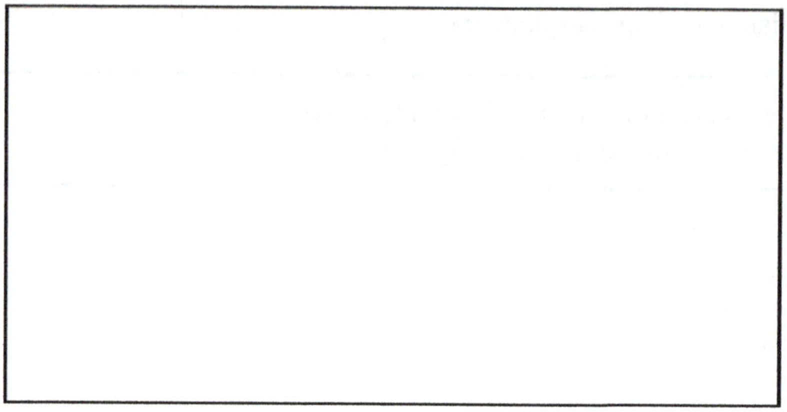

7. Here you see a bunch of quadrilaterals. Label each one. *One* of them does not have a special name.

A ____________________________________

B ____________________________________

C ____________________________________

D ____________________________________

E ____________________________________

F ____________________________________

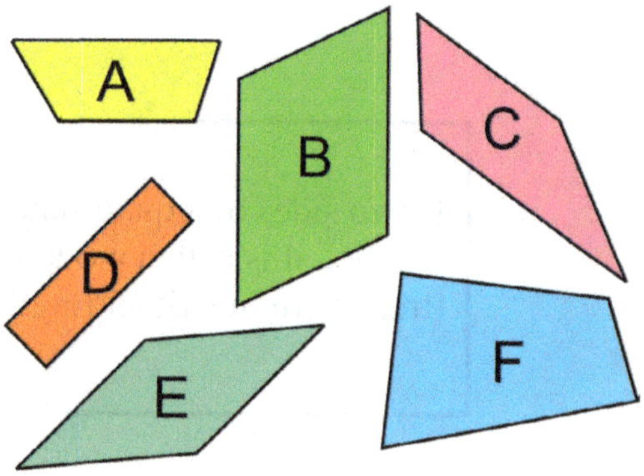

New Terms		
• *parallelogram*	• *rhombus*	• *trapezoid*

Puzzle Corner

Is this rectangle a parallelogram?

How about this square?

And, is a square a rhombus?

How about the rectangle above?

Triangles

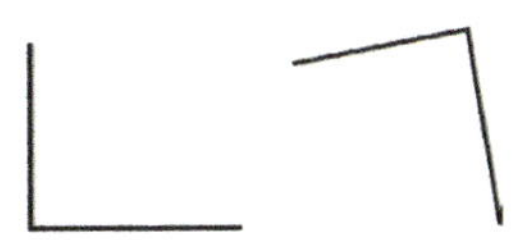

right angles
(exactly 90°)

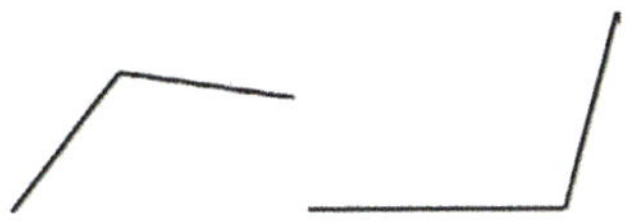

obtuse angles
(more than 90°, less than 180°)

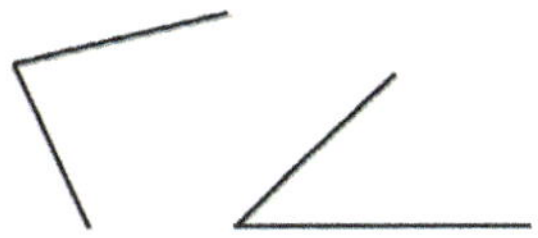

acute angles
(less than 90°)

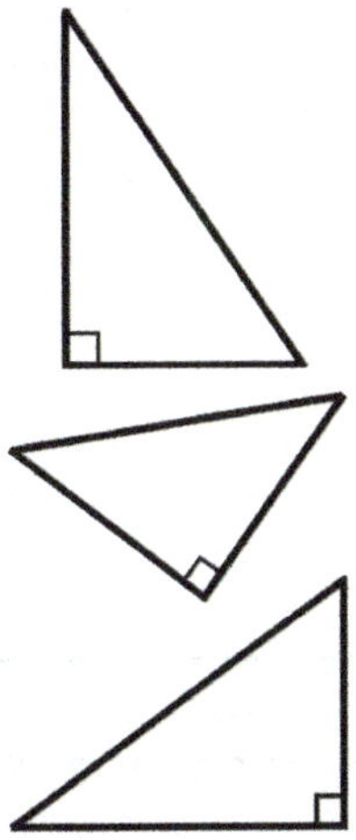

Right triangles have exactly one right angle.

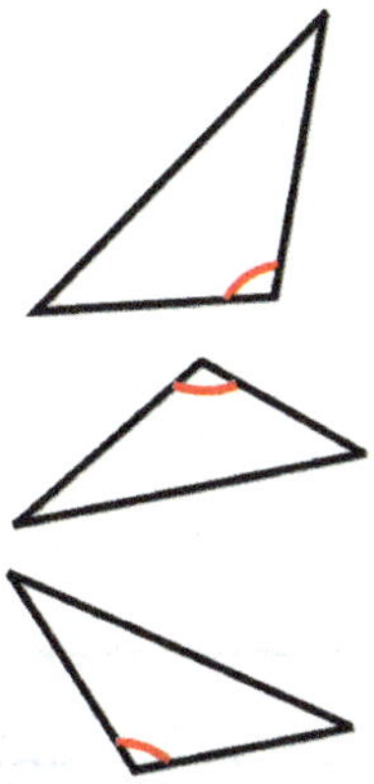

Obtuse triangles have exactly one obtuse angle.

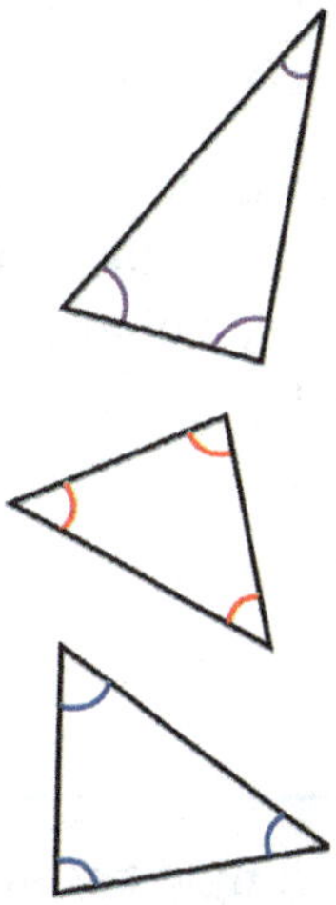

Acute triangles have three acute angles. In other words, *all* the angles are acute.

1. **a.** Draw a right *angle*. Then make it into a right *triangle* by drawing in the third side.

   b. Draw another, different right triangle.

   c. A right triangle has one right angle. Are the other two angles in a right triangle acute, right, or obtuse?

A right triangle has one right angle. The other two angles are _________________.

2. **a.** Draw an obtuse angle. Then make it into an obtuse triangle by drawing in the third side.

b. Draw another, different obtuse triangle.

c. An obtuse triangle has one obtuse angle. Are the other two angles in an obtuse triangle acute, right, or obtuse?

An obtuse triangle has one obtuse angle. The other two angles are ____________________.

3. **a.** Draw any acute triangle.

b. Measure its angles.

They measure _______°, _______°, and _______°.

4. Observe all that you have done in this lesson thus far, and fill in the blanks below.

Right triangles have exactly one __________________ _________________,
and the other two angles are __________________ .

Obtuse triangles have exactly one __________________ _________________,
and the other two angles are __________________ .

Acute triangles have ___ _____________ angles.

5. Label the triangles in the pictures as right, acute, or obtuse.

a.

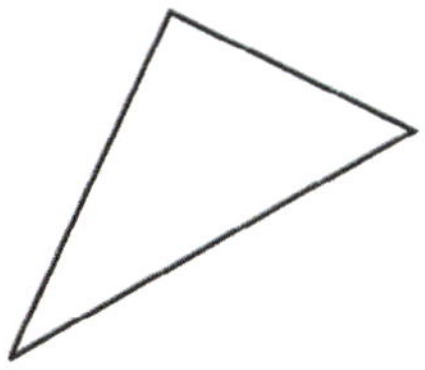

b.

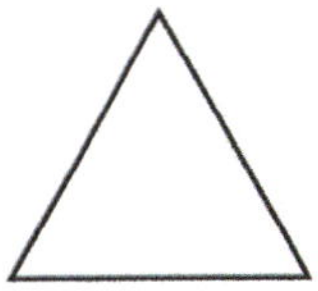

c.

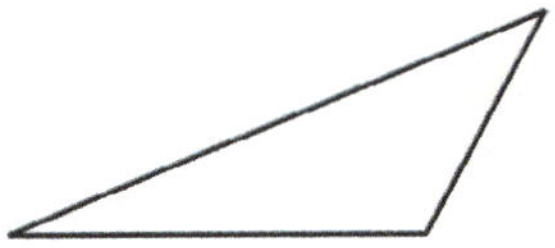

d.

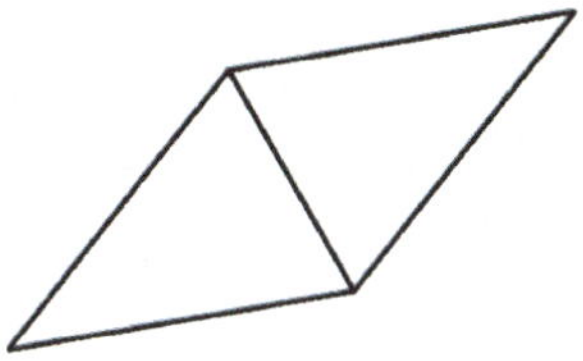

e.

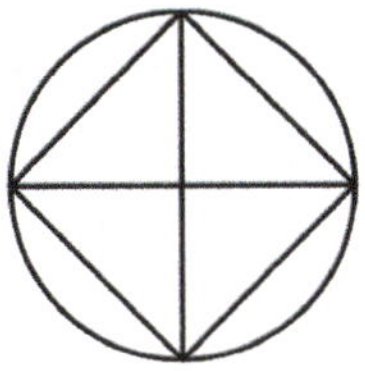

f.

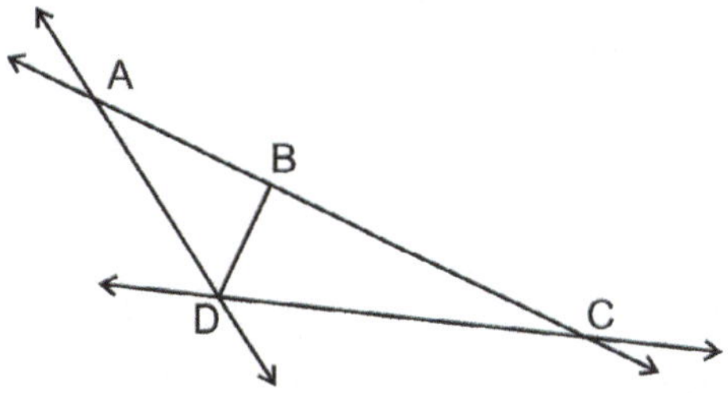

g. triangle ABD:

triangle ACD:

triangle BCD:

6. Label the triangles in the pictures as right, acute, or obtuse.

a. 

National Museum of Marine Corps
Photo by Ron Cogswell

____

b. 

The City Center of Woburn, MA
Photo by Randy Robertson

____

c. 

The White House
Photo by Glyn Lower

____

d. 

Waterlily
Photo by Francis Chung

____

e. 

Warning of a steep hill ahead
Photo by www.flickr.com/photos/shirokazan/

The black triangle is ________________________

The red triangle is ________________________

Find acute, obtuse, and right triangles outside in nature, in buildings, in signs, and so on!

7. **a.** Draw a triangle with 35° and 40° angles. The 35° angle is already drawn for you.

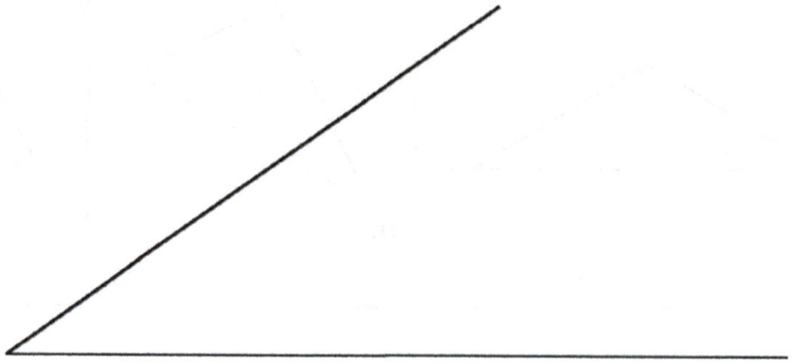

b. Measure the third angle. It is _______ degrees.

c. What kind of triangle is it? (acute, right, obtuse)

8. **a.** Draw a triangle with 125° and 40° angles.

b. Measure the third angle. It is _______ degrees.

c. What kind of triangle is it? (acute, right, obtuse)

9. **a.** Draw a triangle with 55° and 35° angles.

b. Measure the third angle. It is _______ degrees.

c. What kind of triangle is it? (acute, right, obtuse)

New Terms	
• *an acute triangle*	• *an obtuse triangle*
• *a right triangle*	

Line Symmetry

These figures are **symmetrical** in relation to the dashed line. The line is called **a symmetry line**.

Imagine that you FOLDED the figure along the symmetry line. Then both sides would exactly meet. Or, if you placed a mirror along the symmetry line, you would see the other half of the figure reflected in the mirror.

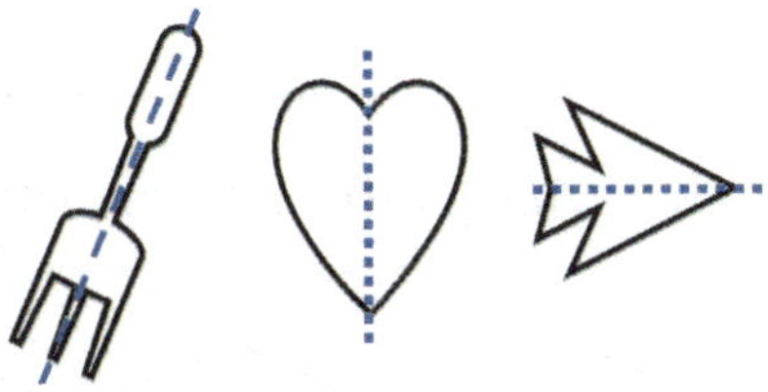

Many figures are not symmetrical at all. You cannot draw a symmetry line in them.

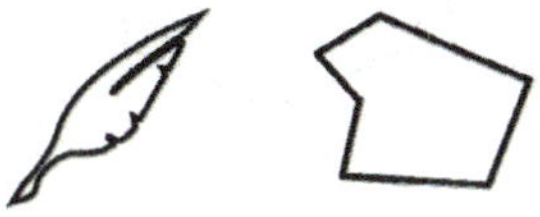

1. Is the line drawn a symmetry line for the figure? You can cut out the images and fold them along the dashed line to check.

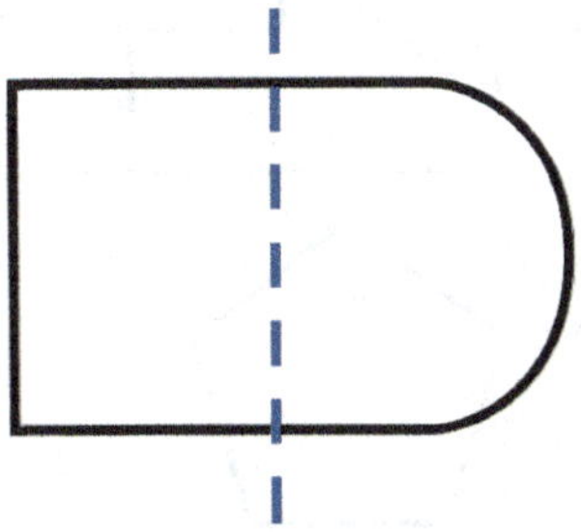

a.

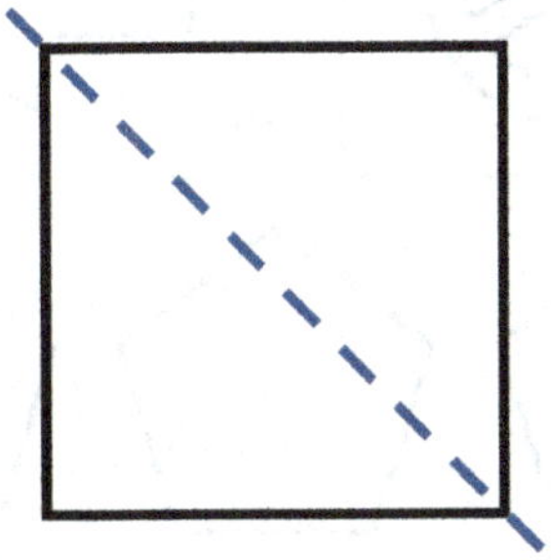

b.

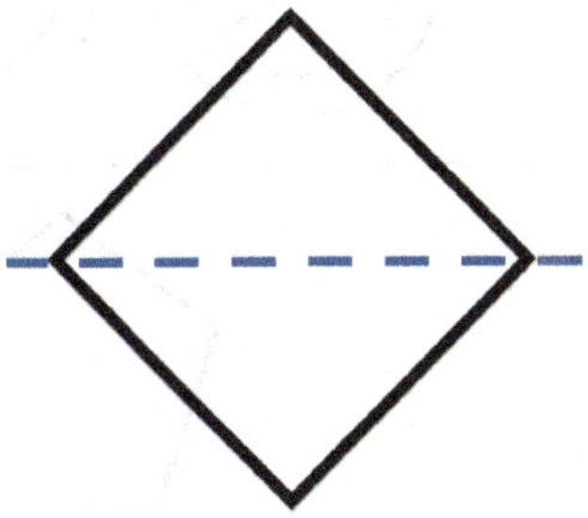

c.

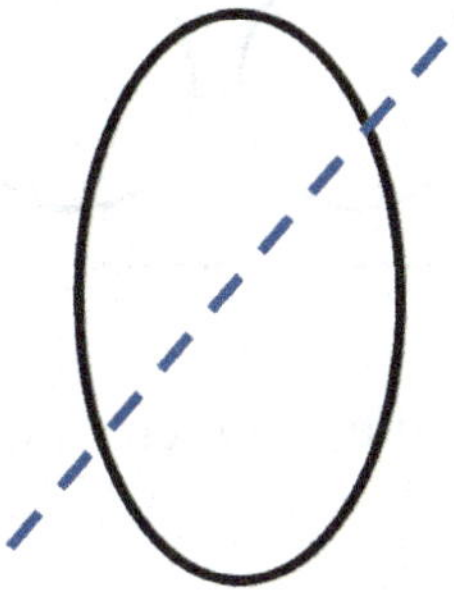

d.

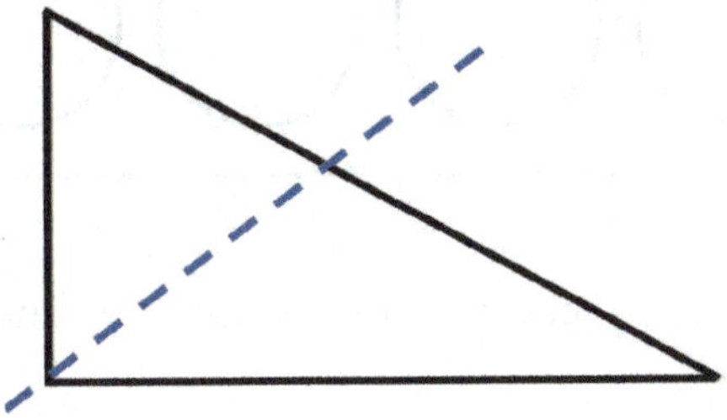

e.

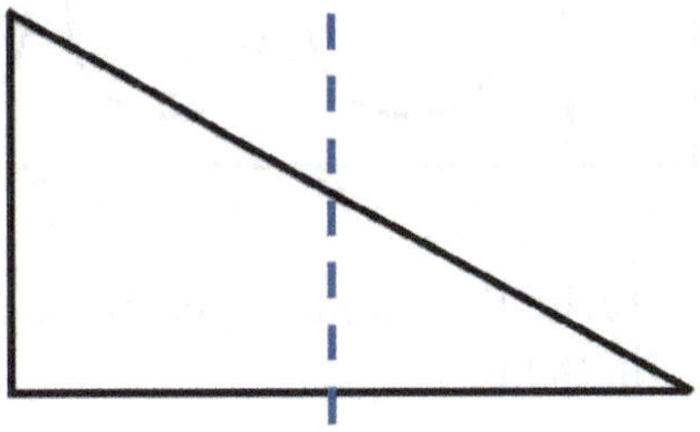

f.

g.

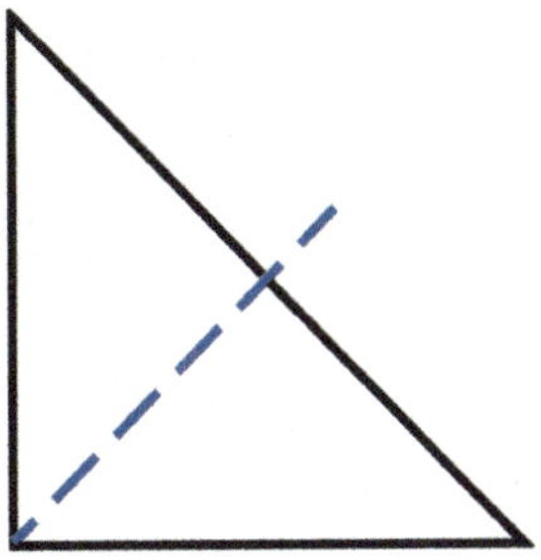

h.

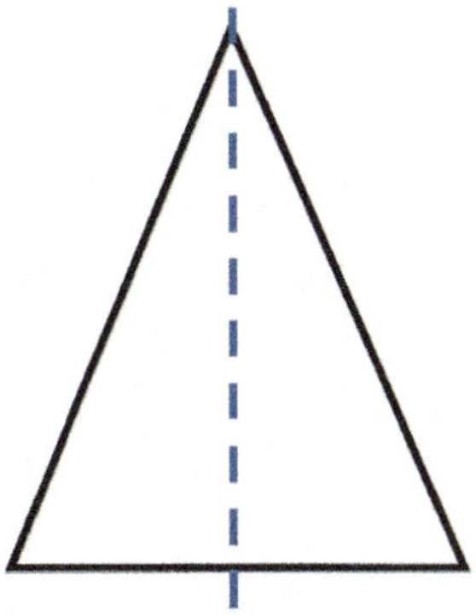

i.

Some shapes can be folded in two different ways so that the sides meet. The cross-shape on the right has *two* different symmetry lines.

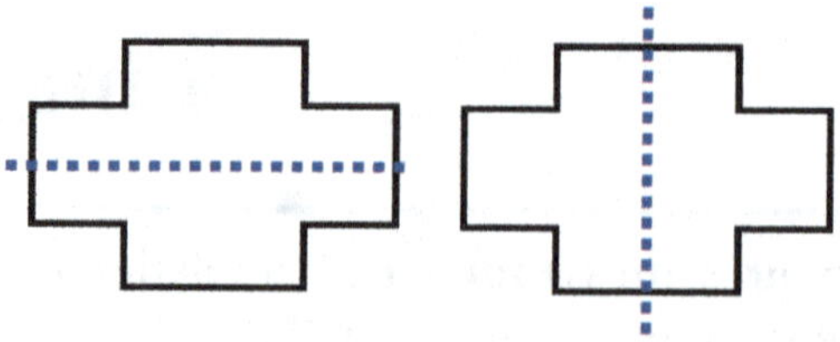

2. Draw as many different symmetry lines as you can in these shapes.

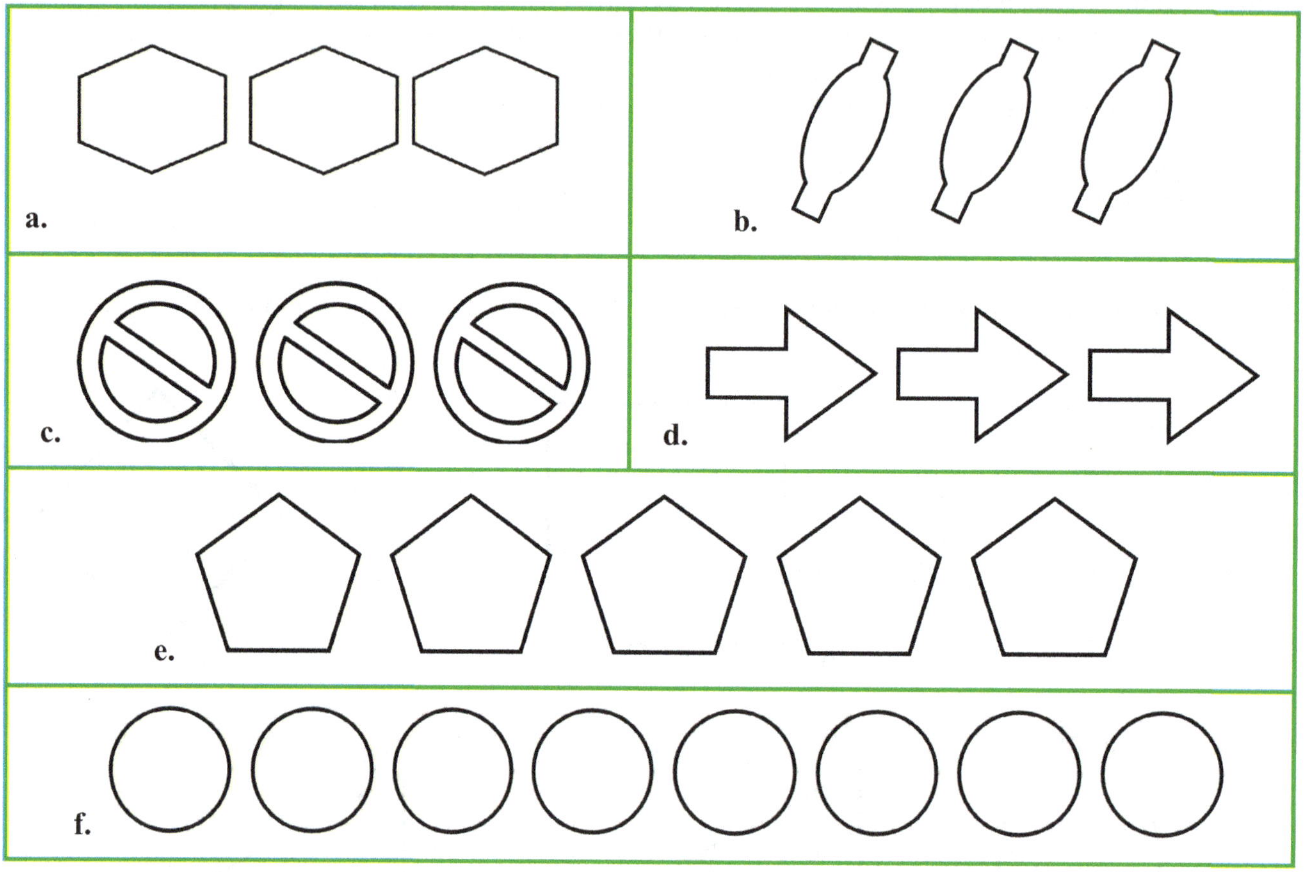

3. Write the capital letters through which you can draw a symmetry line. Draw the symmetry lines in them.

4. Draw a mirror image in the symmetry line to get a symmetrical figure.

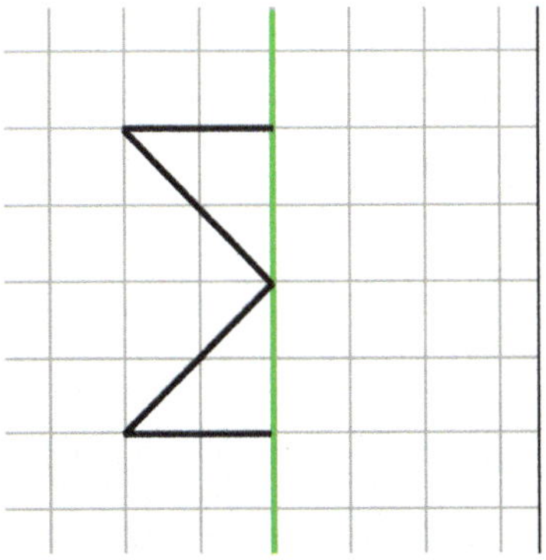

a.

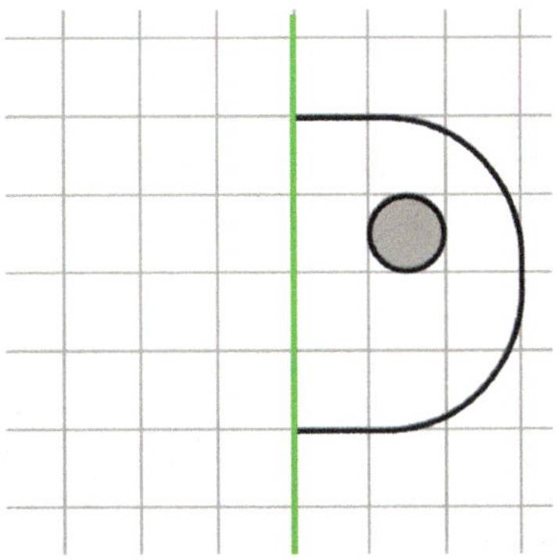

b.

c.

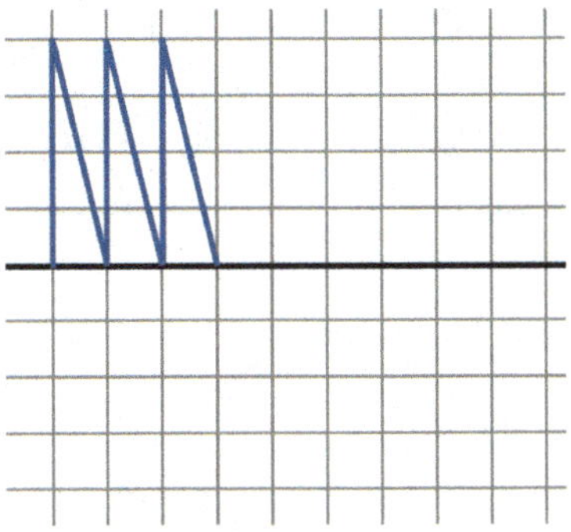

d. Continue the pattern. Then draw its mirror image.

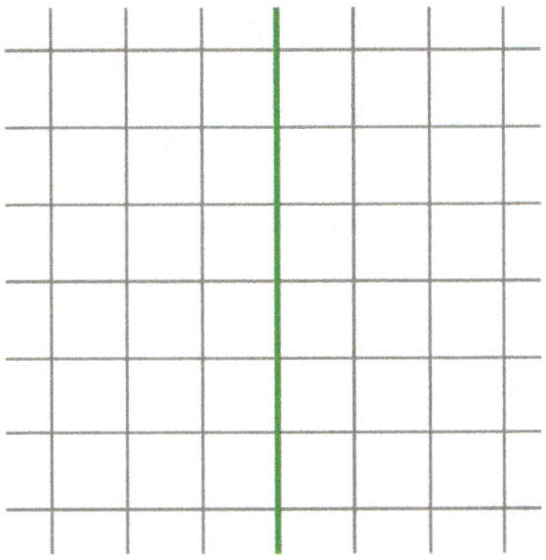

e. Draw your own design and find its mirror image.

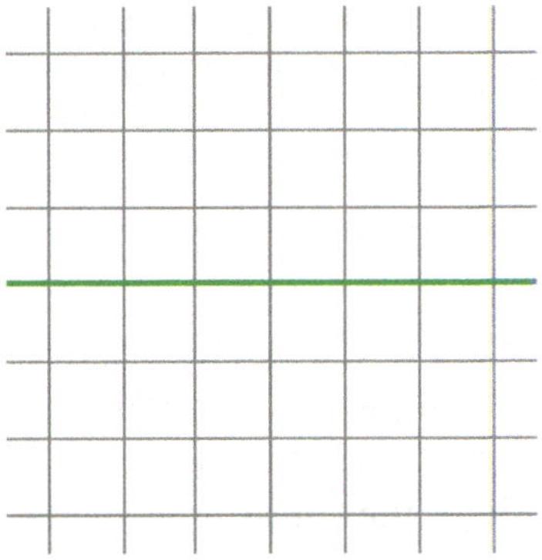

f. Draw your own design and find its mirror image.

5. Look for logos on food products, cars, stores, magazines, and so on. Find at least three logos that have symmetry. Sketch them below. Answer the questions for each logo.

   a. Does the logo employ a square, a rectangle, a triangle, a circle, or some other basic geometric figure in some way?

   b. Does it have any symmetry?

New Terms & Symbols	
• *symmetrical*	• *symmetry line*

Review: Angles

An angle is a figure formed by two rays that have the same beginning point. That point is called the **vertex**. The two rays are called the **sides** of the angle.

Imagine the two sides as being like two sticks that open up a certain amount. The more they open, the bigger the angle.

An angle can be named (1) after the vertex point, (2) with a letter inside the angle, or (3) using one point on the ray, the vertex point, and one point on the other ray.

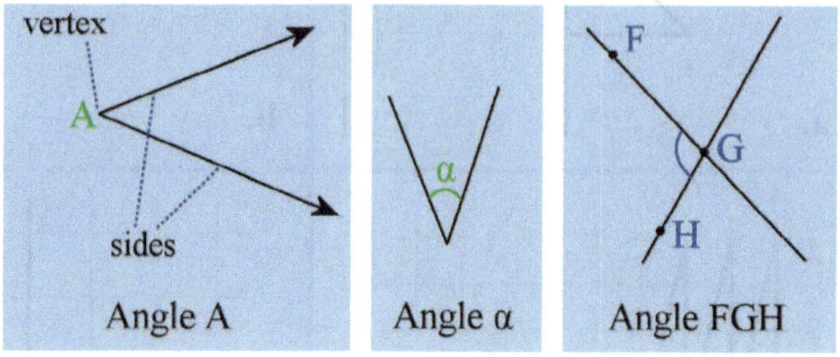

We use **a protractor** to measure angles. The vertex of the angle has to be placed in the middle of the protractor, and ONE side of the angle has to line up with the "zero line" of the protractor.

The angle on the right measures 35 degrees.

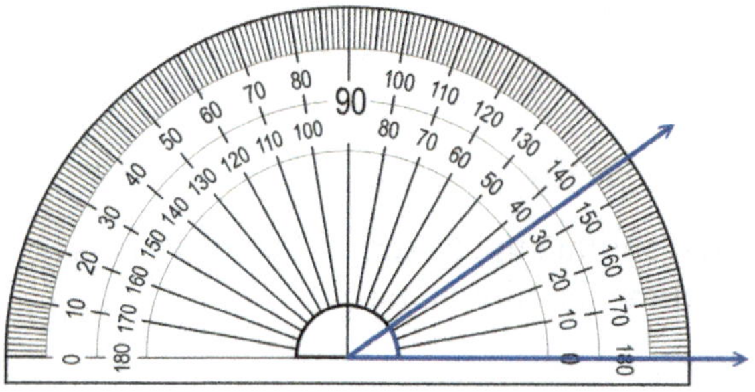

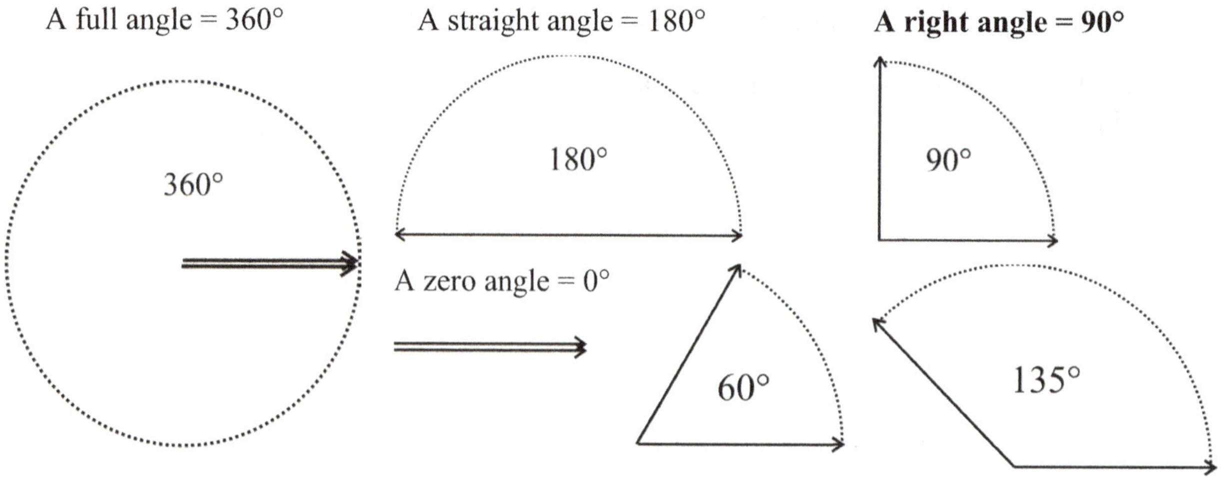

Angles that are more than 0° but less than 90° are called **acute** ("sharp") angles.

Angles that are more than 90° but less than 180° are called **obtuse** ("dull") angles.

Angles that are more than 180° but less than 360° are called *reflex* angles.

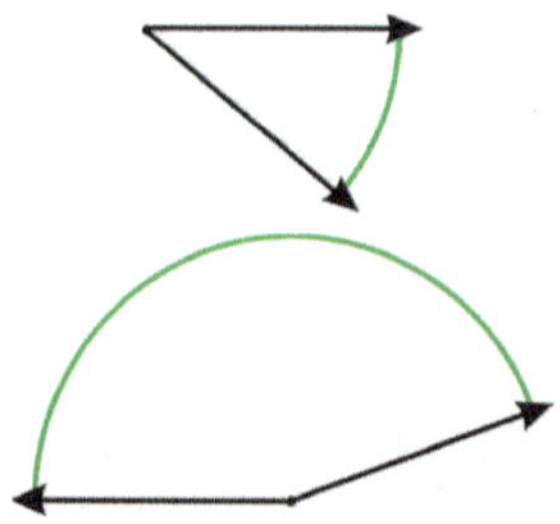

1. Measure these angles with a protractor. If necessary, continue the sides of the angle with a ruler.

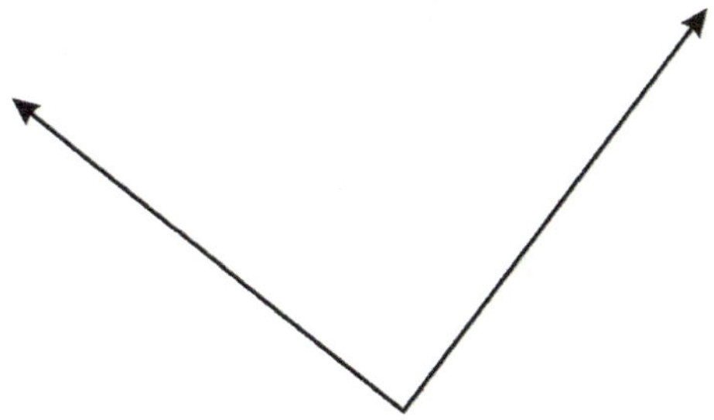

a. ________°

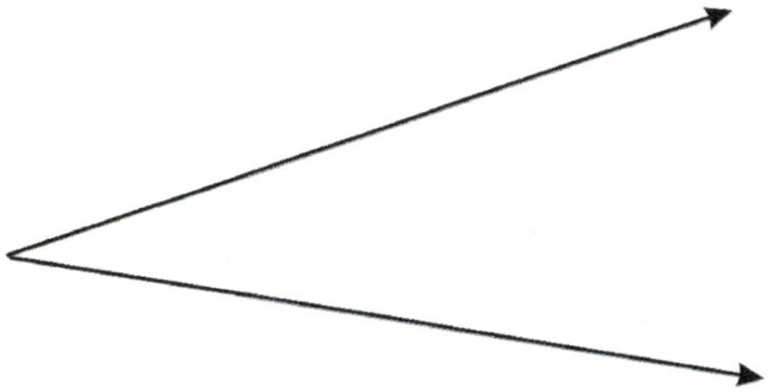

b. ________°

2. **a.** Draw any acute angle, and measure it.

b. Draw any obtuse angle, and measure it.

3. Draw three dots on a blank paper and join them to form a triangle.
Draw the dots far enough apart so that the triangle nearly fills the page.
Then, measure the angles of your triangle.

The angles of my triangle are: ________°, ________°, and ________°.

Classify each angle as acute, right, or obtuse.

4. You see a line and a point on it. The point will be the vertex of an angle. Draw the other side of the angle from the vertex so that the angle measures 76°. Use a protractor.

5. Follow the procedure above to draw acute angles with the following measures:
a. 30° **b.** 60° **c.** 45°

6. Draw obtuse angles with these measures:
a. 135° **b.** 100° **c.** 150°

7. *Estimate* the measure of these angles. Measure to check (you may need to continue the sides).

a.

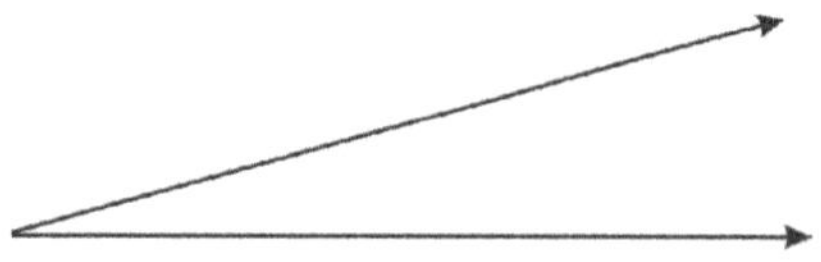

Estimate: ______° Measured: ______°

b.

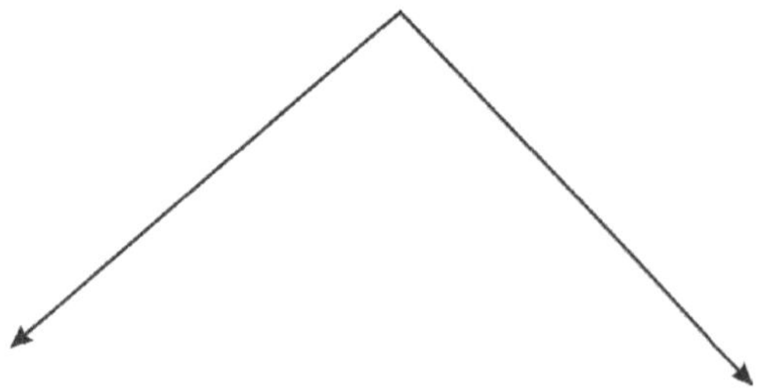

Estimate: ______° Measured: ______°

c.

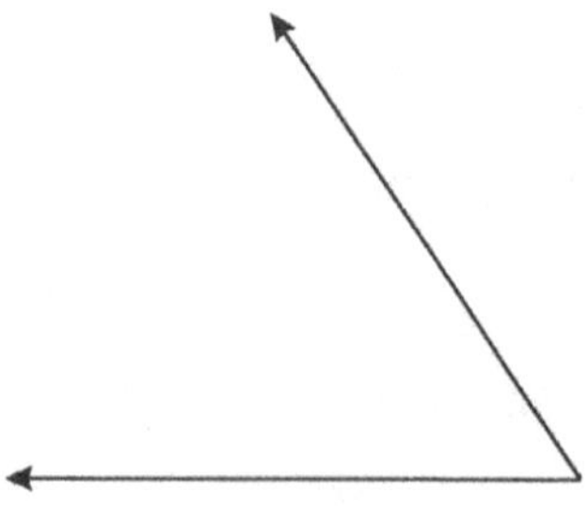

Estimate: ______° Measured: ______°

d.

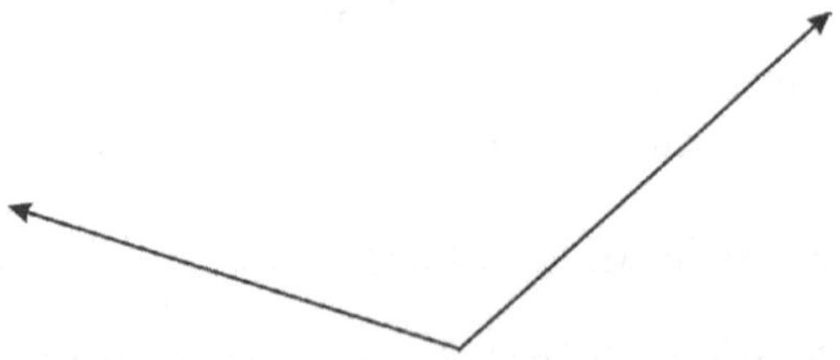

Estimate: ______° Measured: ______°

How to draw a triangle with two given angle measurements (optional)

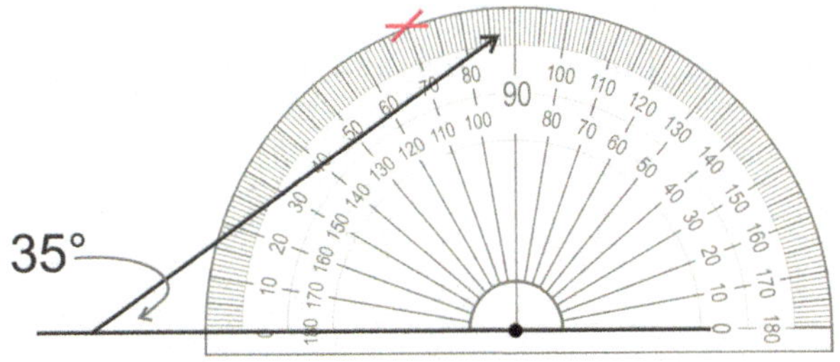

Let's say you have already drawn a 35° angle, and the second angle is supposed to be 70°. The image shows you how to place your protractor so you can measure and mark the 70° angle.

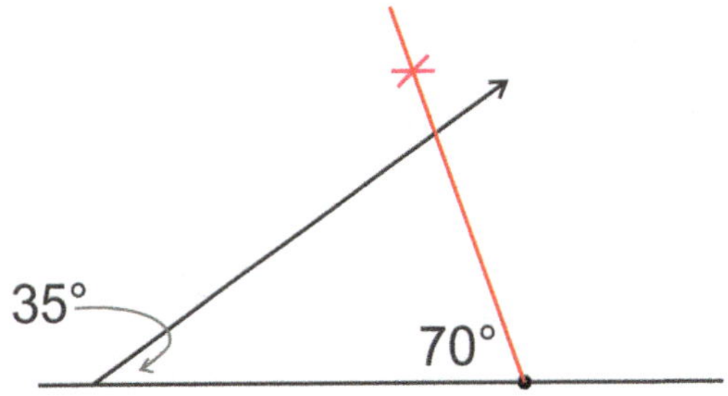

Then remove the protractor and draw the third side of the triangle.

8. (optional)

   a. Draw a triangle with 50° and 75° angles. It can be of any size — smaller or bigger.
   Hint: Start out by drawing a (long) horizontal line, and two dots on it which mark the two vertices of the triangle.

   b. Measure the third angle. It measures _______°.

   c. Label each angle in the triangle as acute, obtuse, or right.

9. (optional)

   a. Draw a triangle with 110° and 35° angles.

   b. Measure the third angle. It measures _______°.

   c. Label each angle in the triangle as acute, obtuse, or right.

Polygons

A **polygon** is a flat, two-dimensional figure that consists of line segments, and is *closed.*

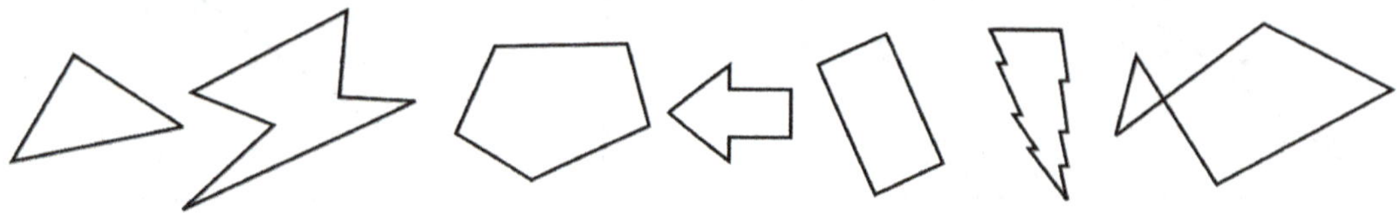

The boundary of a polygon is allowed to cross itself, like in the polygon above at the right. However, in this chapter we will mostly deal with *simple* polygons where such does not happen.

These figures are not polygons. Notice how each figure either is not closed, does not consist of line segments, or is not a flat, two-dimensional figure:

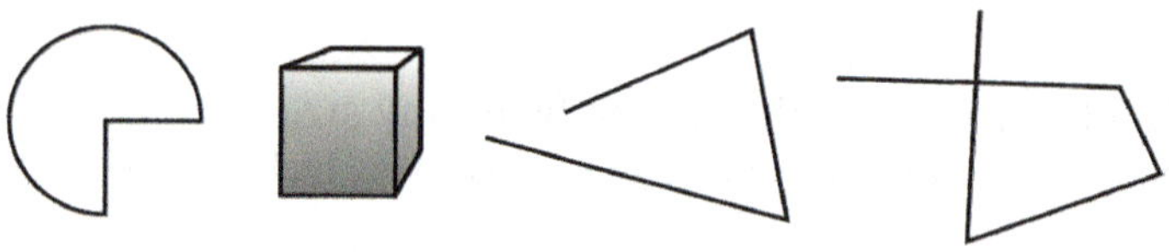

Polygons are named after the number of vertices they have. Most of the names for polygons in English have their roots in Greek, using a number and the Greek word "*gonia*" which means "angle".

Vertices	Name	Greek/Latin
3	triangle	tri = three
4	quadrilateral	quadri (Latin) = four
5	pentagon	pente = five
6	hexagon	hex = six
7	heptagon	hepta = seven
8	octagon	okto = eight

1. Classify each figure as a polygon, or not a polygon.

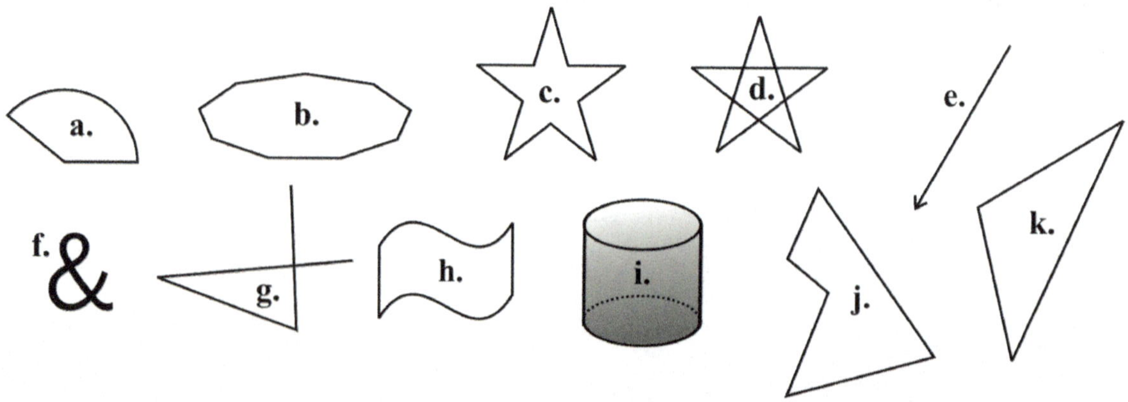

A **regular polygon** is one with congruent sides and angles. In other words, the sides are of equal length, and the angles have the same measure.

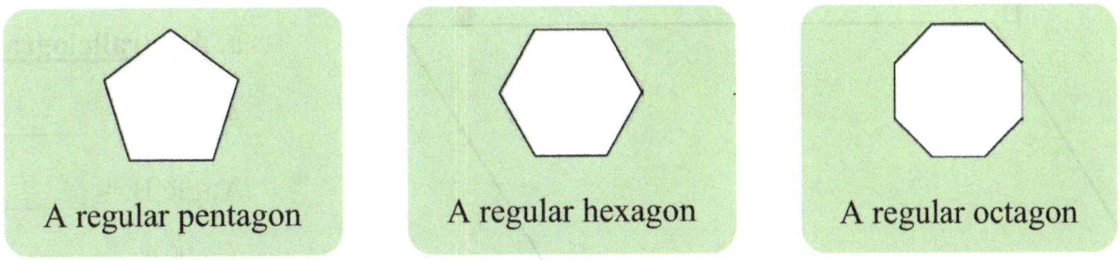

2. What is the common name for a *regular quadrilateral*?

3. Name the figures. Include the word "regular" when fitting.

a. ______________________________

b. ______________________________

c. ______________________________

d. ______________________________

e. ______________________________

f. ______________________________

g. ______________________________

h. ______________________________

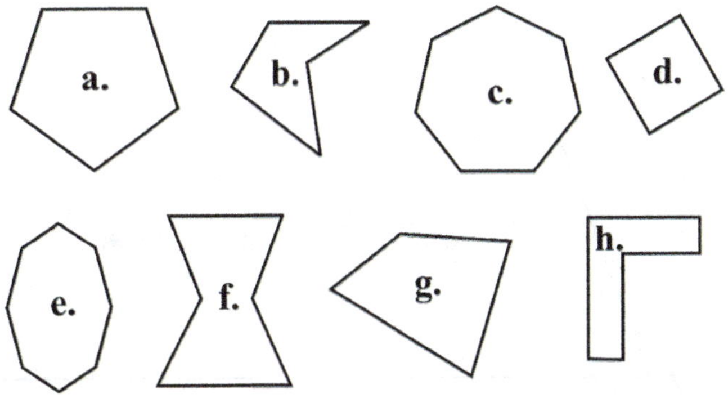

4. For each of the regular octagon and regular pentagon below, measure one of the angles to find the measure of *all* the angles in it (since it is *regular*!).

a.

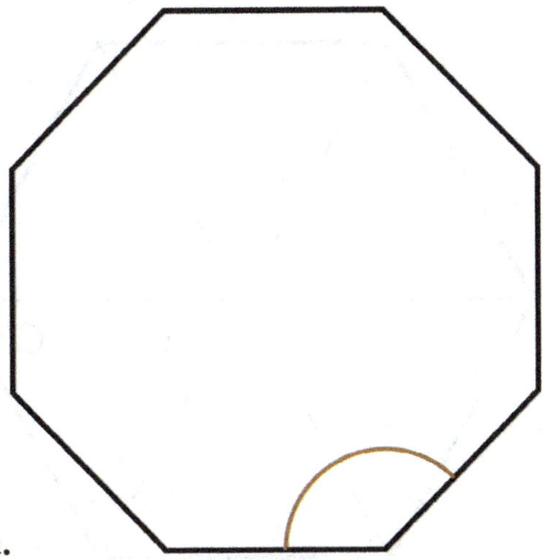

b.

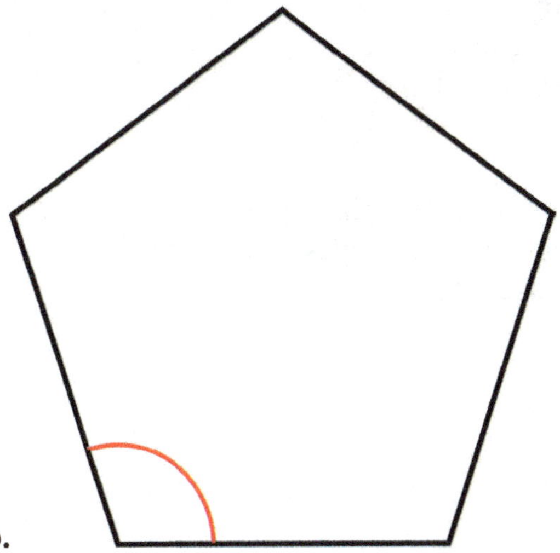

5. Find the measures of all the angles in these quadrilaterals. What do you notice?

a. A parallelogram

Angle A = ________°

Angle B = ________°

Angle C = ________°

Angle D = ________°

I notice that ____.

b. A trapezoid

Angle A = ________°

Angle B = ________°

Angle C = ________°

Angle D = ________°

I notice that ____.

Puzzle Corner

This is a regular hexagon.
Figure out the angles A, B, and C.
Measure to check.

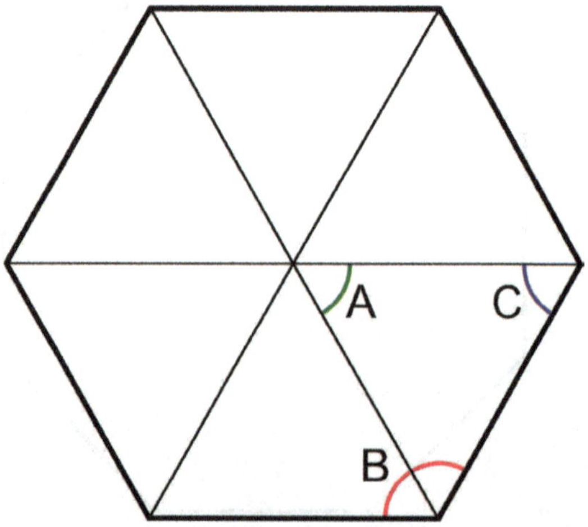

Classifying Quadrilaterals 1

Quadrilaterals are polygons with four sides (*quadri-* = four, *lateral* = referring to a side).

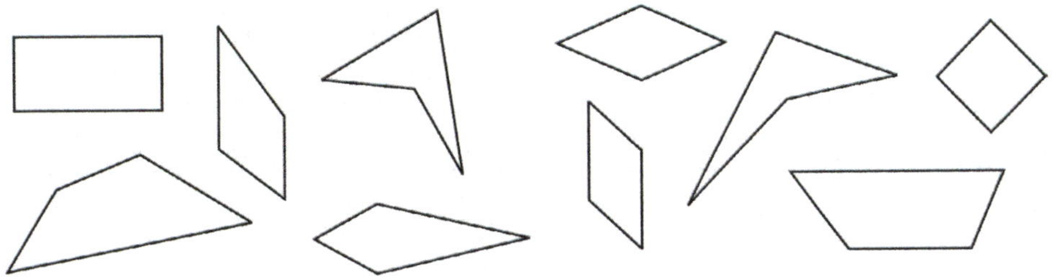

There are many different kinds of quadrilaterals. We could classify, or sort, them in various ways. The sorting could be done on the basis of the types of angles, or on the basis of the lengths of sides — and these characteristics are used. But the main way that we will *start* the classification of quadrilaterals has to do with whether or not the quadrilateral has **any parallel sides**.

1. Sort the quadrilaterals to groups according to the number of parallel sides each has. You can write the letter of each in the box below, or cut them out. Downloadable version here:

https://www.mathmammoth.com/download/sort_quadrilaterals.pdf

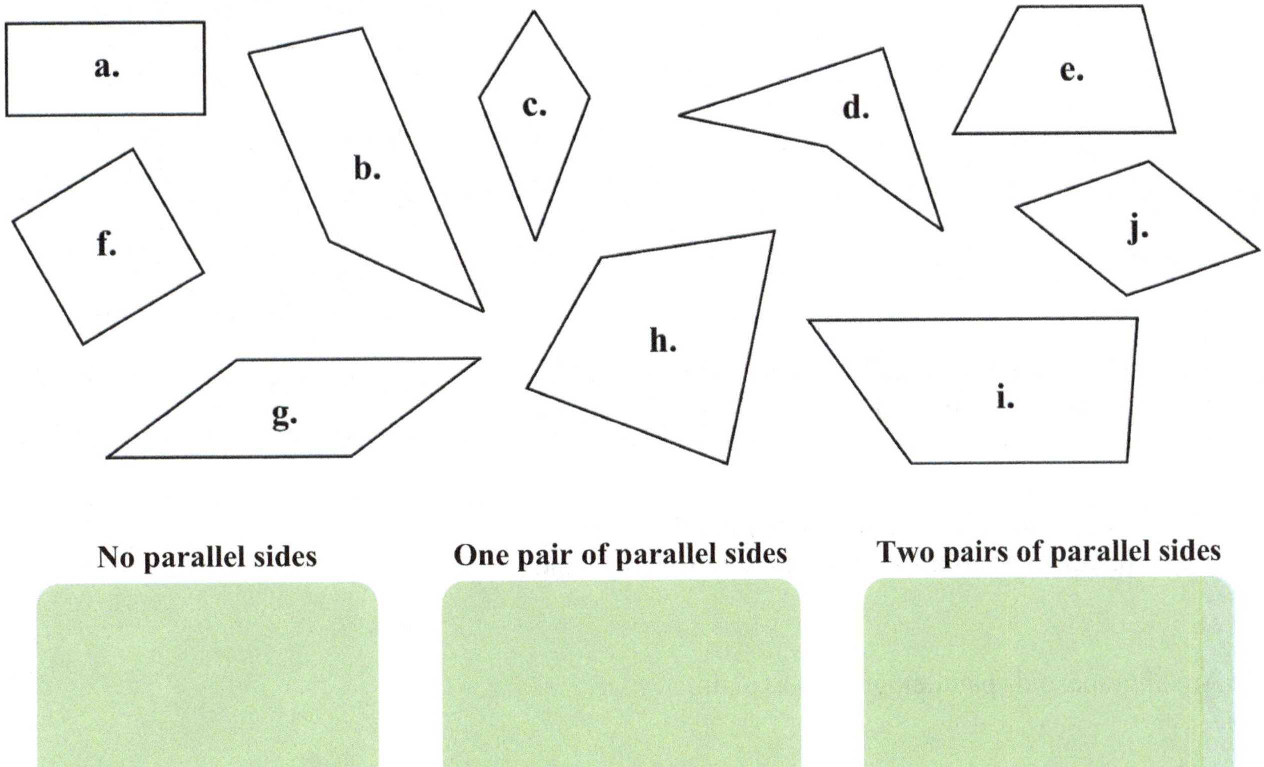

No parallel sides	One pair of parallel sides	Two pairs of parallel sides

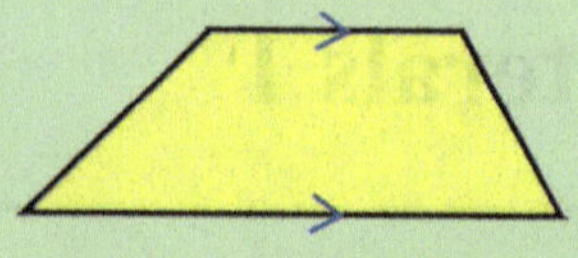

A **trapezoid** has at least one pair of parallel sides.

The arrowheads mark the pair of parallel sides.

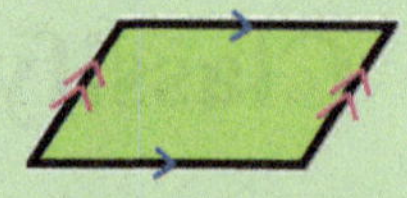

A **parallelogram** has two pairs of parallel sides.

The single arrowheads mark the one pair of parallel sides, and the double arrowheads mark the other.

2. In the grid below:

a. Join the dots to make a parallelogram.

b. Draw one other, differently shaped parallelogram.

c. Draw two different trapezoids.

d. Draw a quadrilateral with no parallel sides.

3. **a.** Is a rectangle always a parallelogram? In other words, does a rectangle fulfill the definition of a parallelogram?

b. Is a parallelogram always a trapezoid? In other words, does a parallelogram fulfill the definition of a trapezoid?

c. Are all trapezoids parallelograms? Explain.

d. Are all rectangles trapezoids? How about vice versa?

On a previous page, you sorted the quadrilaterals, making three separate groups or sets. But, we can also use a Venn diagram, and show how a set might overlap another, or be inside another.

For example, the Venn diagram on the right shows a group of mammals and a group of aquatic animals. But they overlap; some animals are *both* mammals and aquatic.

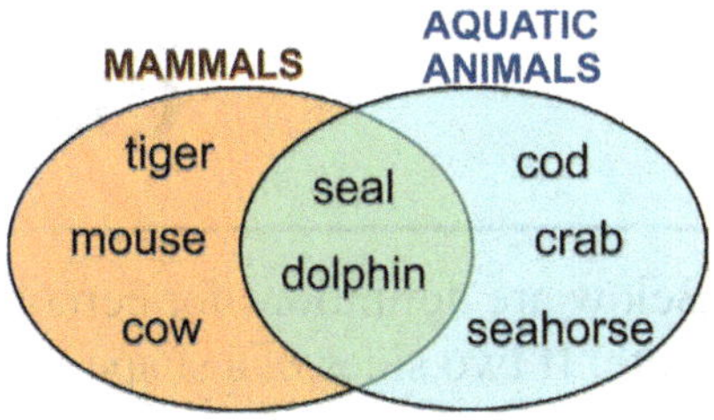

In the case of trapezoids and parallelograms, the one set is inside the other:

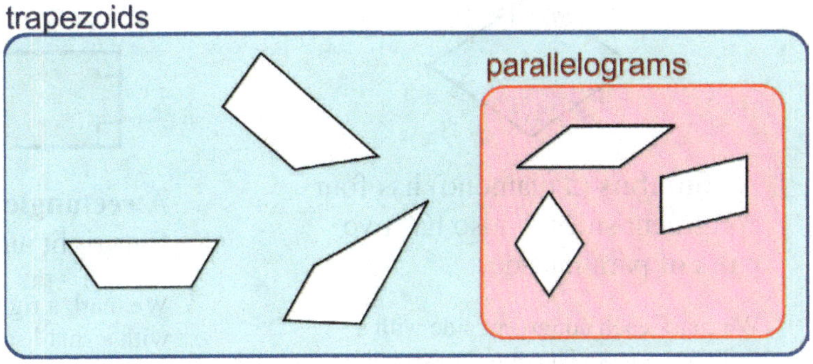

4. Why is the set of parallelograms *inside* the set of trapezoids in the Venn diagram above?

5. Where would rectangles go in the Venn diagram above?

6. Outside the sets for trapezoids and parallelograms, we can draw a third set, which includes all the rest of the quadrilaterals with *no* parallel sides. Sort the shapes below into the correct sets by sketching them in the sets.

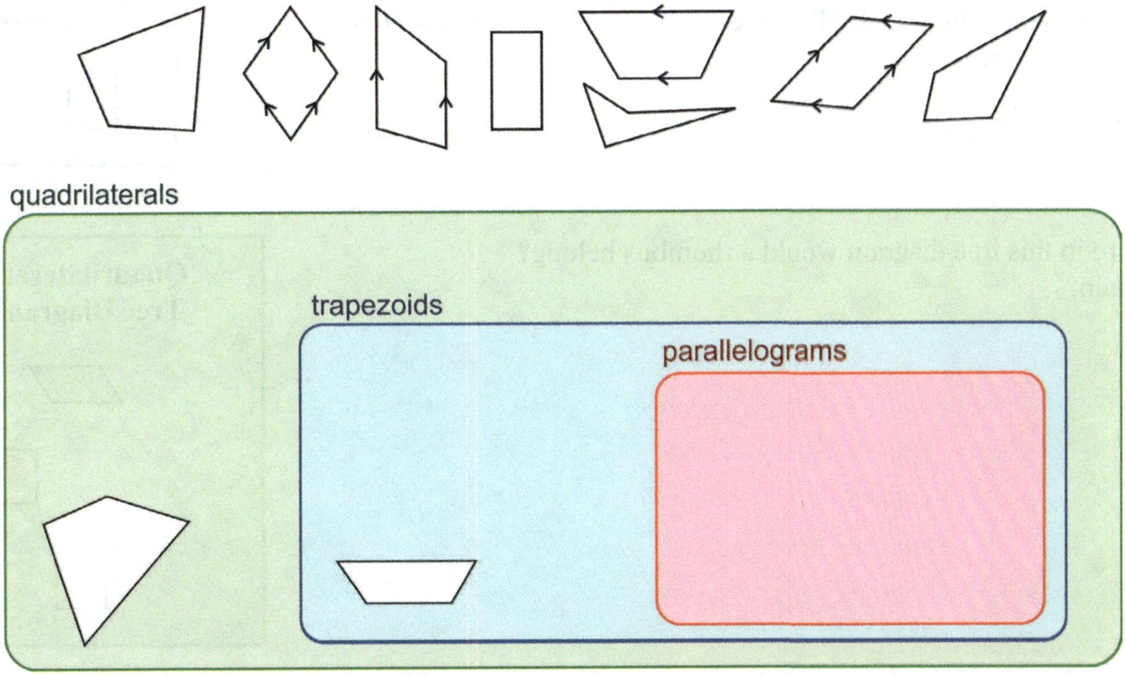

Classifying Quadrilaterals 2

Below are definitions for certain special quadrilaterals. Study them carefully.
Note: If two sides of a shape are **congruent**, it means they have the same length.

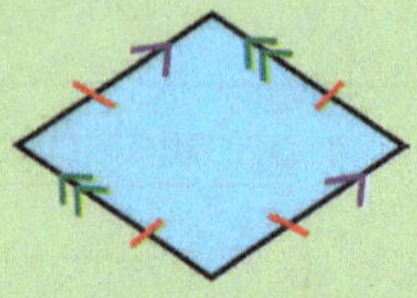

A **rhombus** (a diamond) has four congruent sides. It also has two pairs of parallel sides.

We mark each congruent side with a little tick mark. The plural of rhombus is "rhombi" or "rhombuses".

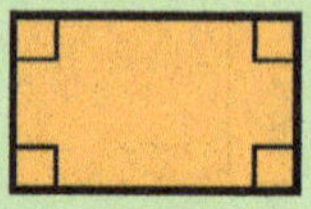

A **rectangle** has four right angles.

We mark a right angle with a small square.

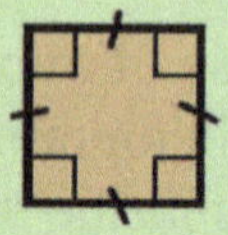

A **square** has four right angles *and* four congruent sides.

1. Does a square fulfill the definition of a rhombus?

2. In the tree diagram on the right, you see a rectangle, a parallelogram, and a square. Start "reading" the tree diagram from the top, beginning with the parallelogram. Fill in:

   a. The ______________ is like a "child" to the parallelogram: it *is* a parallelogram. Additionally, its angles are right angles.

   b. The square is like a "child" to the ______________ : it, too, has four right angles.

   Additionally, ____.

A Tree Diagram

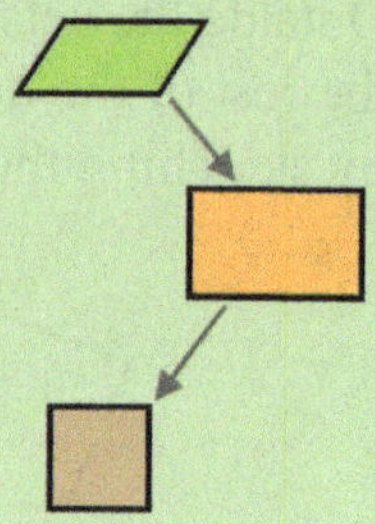

3. Where in this tree diagram would a rhombus belong? Explain.

Quadrilaterals Tree Diagram

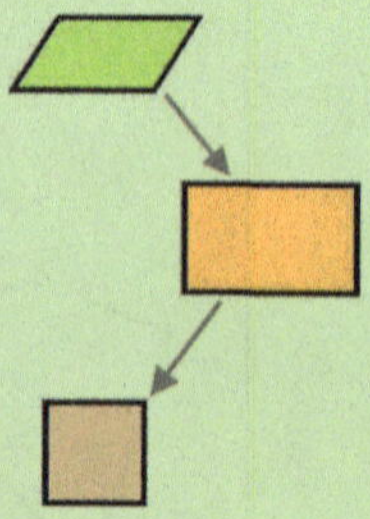

4. Below, you see six shapes. Sketch them in the Venn diagram, and label the two sets, in a manner that makes sense.

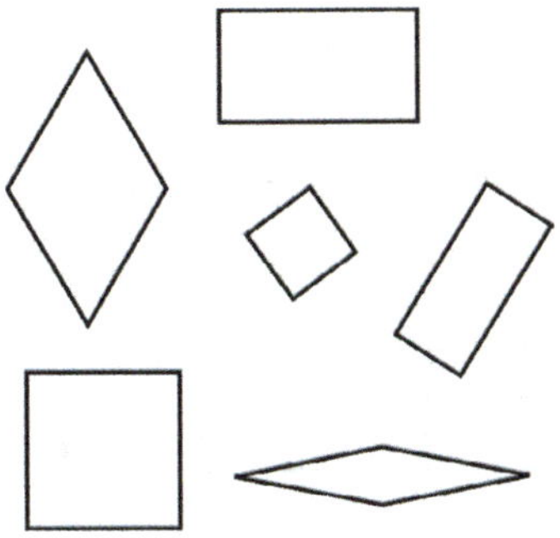

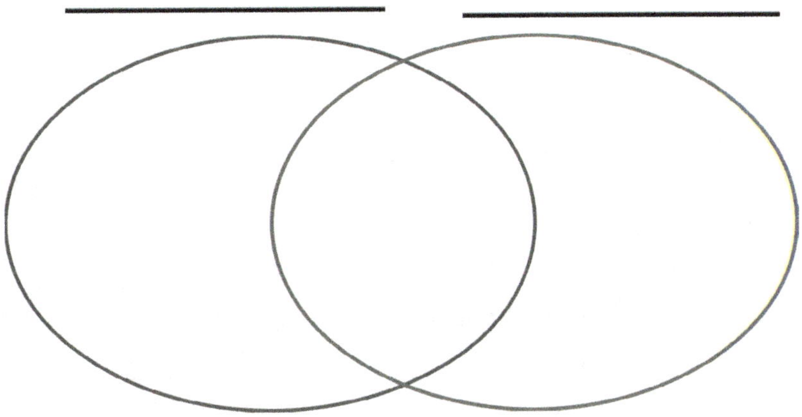

A **kite** is a quadrilateral with two pairs of congruent sides that are **adjacent** (touch each other).

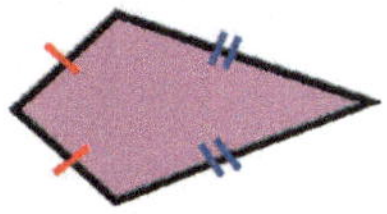

5. Do these quadrilaterals also fulfill the definition of a kite?

a. rhombus

b. rectangle

c. square

d. trapezoid

6. Each quadrilateral below is either a parallelogram, a rhombus, a trapezoid, or a kite. Write their names.

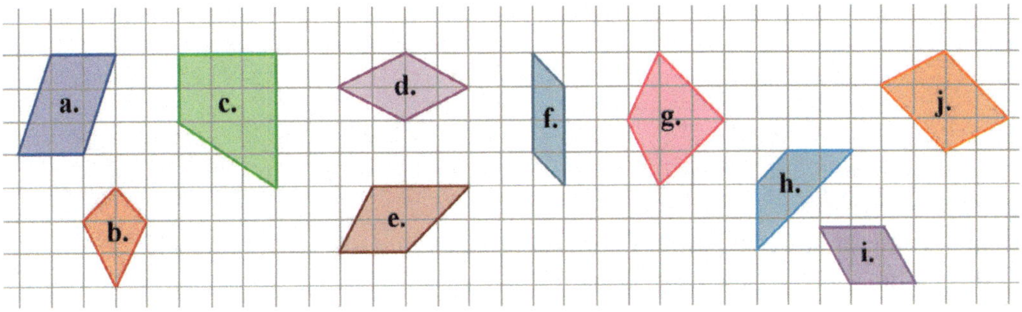

a. ______________________ **b.** ______________________

c. ______________________ **d.** ______________________

e. ______________________ **f.** ______________________

g. ______________________ **h.** ______________________

i. ______________________ **j.** ______________________

7. Below you see our final Venn diagram for quadrilaterals. Sketch the seven shapes below in the proper places in the diagram.

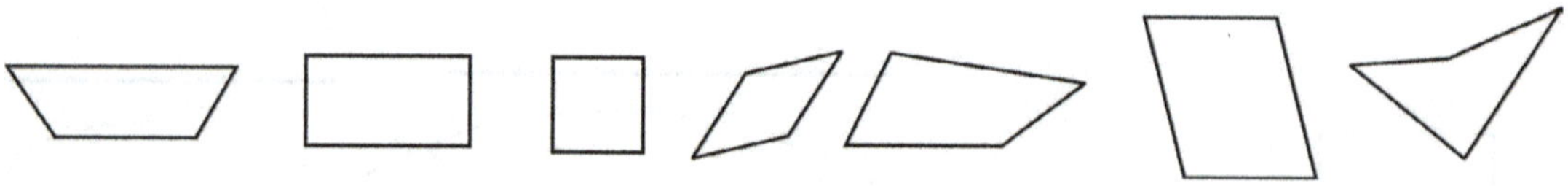

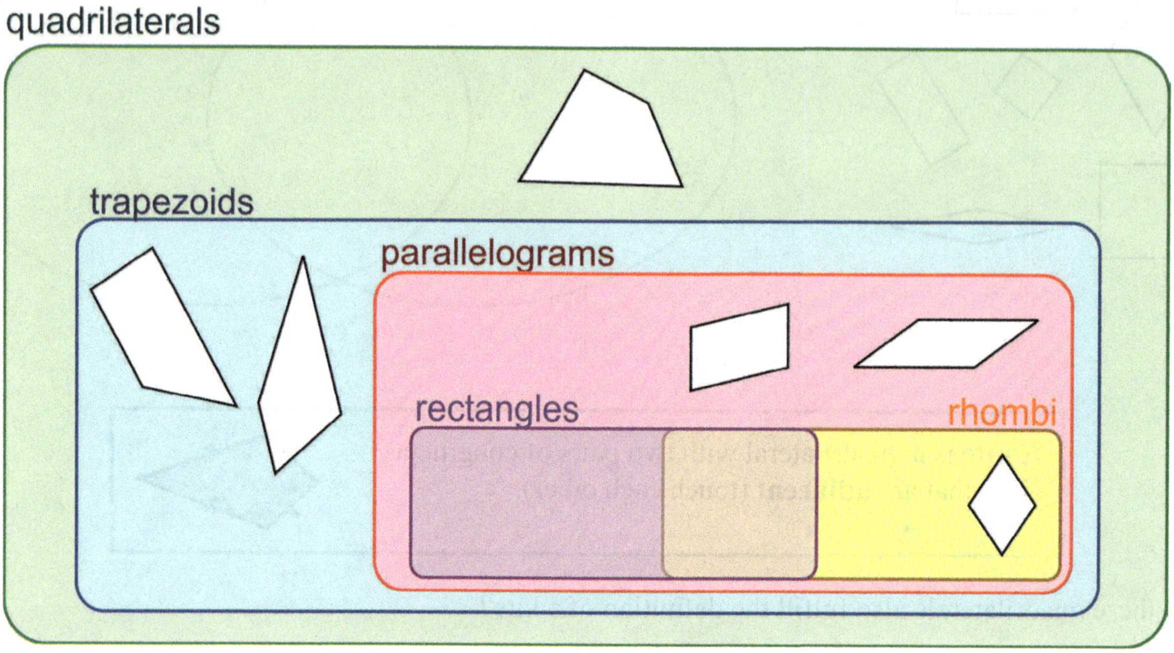

8. Jeanine is supposed to draw a quadrilateral with each side 1½ inches long. Can she draw only one kind, or can she draw several different-looking ones? Explain.

9. A certain quadrilateral has two pairs of congruent sides *and* also two pairs of parallel sides. What kind of quadrilateral is it?

10. The sides of a trapezoid measure 5 cm, 3 cm, 8 cm, and 3 cm.
    Is it also a kite? A parallelogram? A rhombus?

Classifying Quadrilaterals 3

(This lesson is optional.)

1. Consider a quadrilateral that has **four right angles** and **one side 2 inches long**. Draw sketches of such quadrilaterals on blank paper. (A rough sketch is enough; we are not looking for accurate drawings here.)

   Can you draw only *one* such quadrilateral, or *several* different ones that all have a different shape and/or size?

2. The sides of a parallelogram measure 2 in, 3 in, 2 in, and 3 in.
   Is it also a kite?   A rhombus?   A trapezoid?

3. Draw a trapezoid so that its parallel sides measure 3 in and 1 ¾ in.
   Can you draw several such trapezoids that are not identical?

In a **scalene** quadrilateral, all sides are of different lengths (no two sides are congruent).

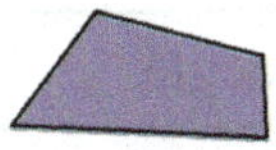

4. Here is a complete tree diagram for quadrilaterals! The top figure signifies a generic quadrilateral. When a shape is under another, it signifies it belongs to that "family".

   Name each type of quadrilateral in the diagram.

Quadrilaterals Tree Diagram

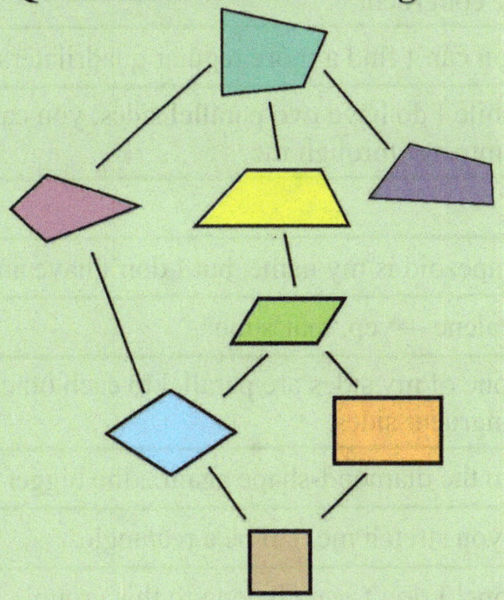

5. Answer the questions. The tree diagram will help.

   a. Is a rhombus also a kite?

   b. Is a square also a kite?

   c. Is a rectangle also a kite?

   d. Is a square also a trapezoid?

   e. Is a parallelogram also a kite?

6. (*Challenge*) Solve the quadrilaterals puzzle and uncover a message! One shape may correspond to two letters.

Quadrilateral	Letter
I have one right angle. I look like a toy that comes back to you.	
I have the biggest area around here!	
I am the little rhombus!	
Two (and only two) of my sides are parallel. The other two are congruent.	
There's nothing congruent or parallel about me.	
I'm the bigger parallelogram with four congruent sides.	
I am the big rhombus!	
I've got right angles and two pairs of congruent sides.	
I have four right angles and four congruent sides!	
Two—and only two—of my sides are parallel. None are congruent.	
They call me a rectangle.	
I'm one of the parallelograms in the bunch with a vowel inside me.	
I have two pairs of parallel sides, but not all of my sides are congruent.	
You can't find a more regular quadrilateral than me.	
While I do have two parallel sides, you cannot draw a line of symmetry through me.	
Trapezoid is my name, but I don't have any congruent sides.	
Scalene—yep, that's me!	
None of my sides are parallel to each other, but I have two pairs of congruent sides.	
I'm the diamond-shape again... the bigger one.	
If you stretch me, I'll be a rectangle.	
Oops! I don't even belong to this group!	
	!

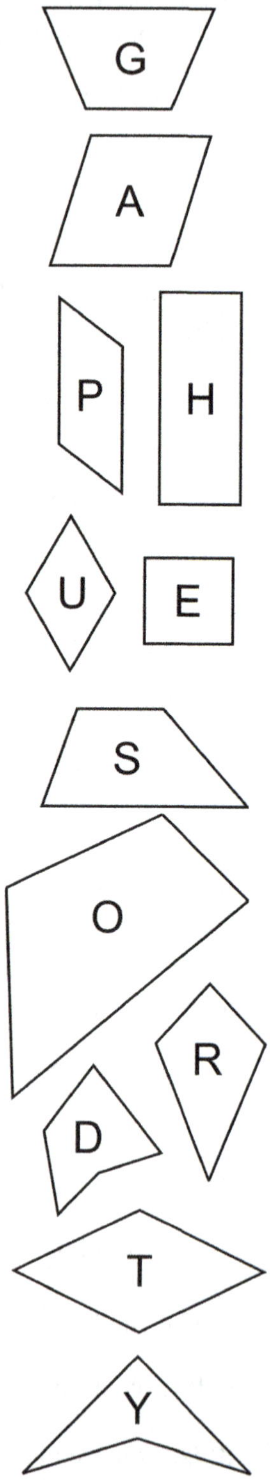

Classifying Triangles 1

We cannot classify triangles based on how many parallel sides they have (why?).

So... what characteristics could we use to classify, or sort, triangles?
Give it some thought in the following exercise.

1. Sort the following triangles according to some characteristics. You can cut these out! You may want to use a ruler and/or a protractor. Downloadable version is here:

   https://www.mathmammoth.com/download/sort_triangles.pdf

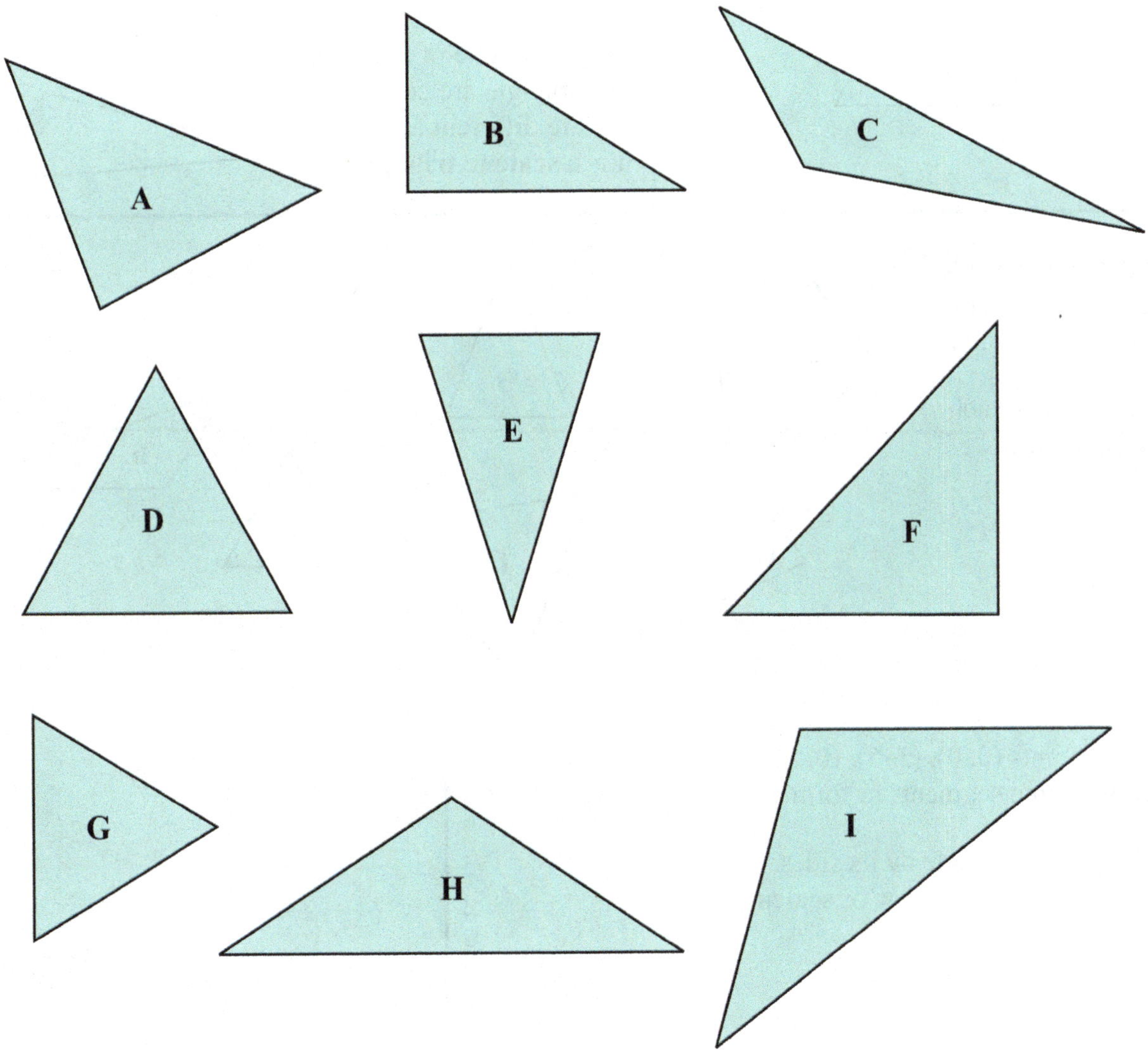

Triangles can be classified either by **side lengths** or by **angles**. We will look at both ways!

Classification according to sides

In an **equilateral triangle**, all three sides of the triangle are congruent.

"*Equi-*" refers to things that are the same or equal, and "*lateral*" refers to sides. Think of it as a "same-sided" triangle.

If only *two* of a triangle's sides are congruent, then it is called an **isosceles triangle**.

Think of it as a "same-legged" triangle, the "legs" being the two sides that are the same length.

MARK the two congruent sides of each isosceles triangle with a tick mark:

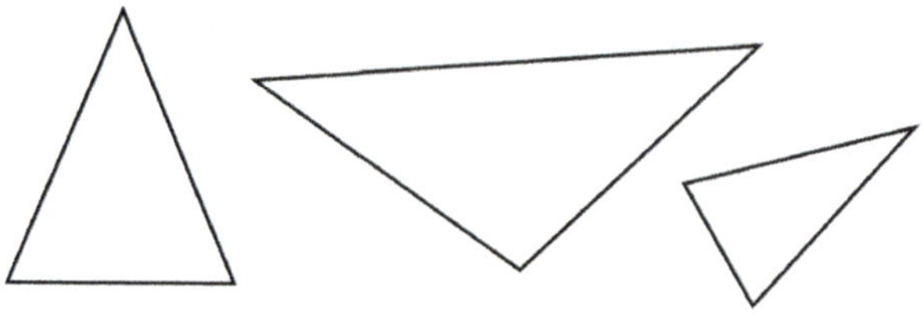

Lastly, if none of the sides of a triangle are congruent (all are different lengths), it is a **scalene triangle**.

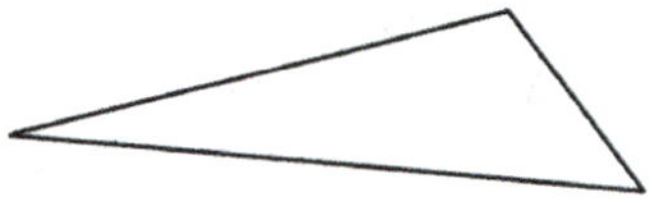

2. Classify the triangles as either equilateral, isosceles, or scalene.

   You can mark each triangle with an "*e*," "*i*," or "*s*" correspondingly.

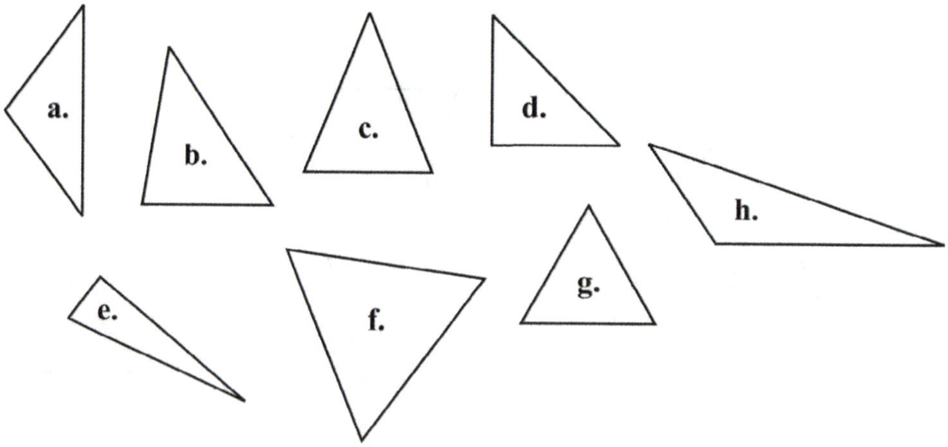

3. Plot the points (0, 0), (3, 5), (0, 5) , and connect them with line segments to form a triangle.

   Classify your triangle by its sides.
   Is it equilateral, isosceles, or scalene?

   ____

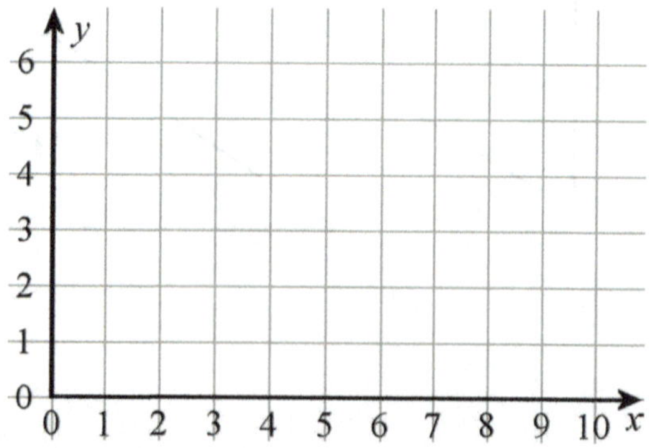

Classification according to angles

We can also classify a triangle according to its angles.

- A **right triangle** has one right angle.
- An **obtuse triangle** has one obtuse angle.
- An **acute triangle** has three acute angles.

Notice that the definition of a *right* triangle only requires *one* of the angles to be a right angle. Similarly, a triangle is obtuse if only *one* of its angles is obtuse. Yet, the definition for an acute triangle mentions *three* acute angles.

4. Classify each triangle as acute, right, or obtuse.

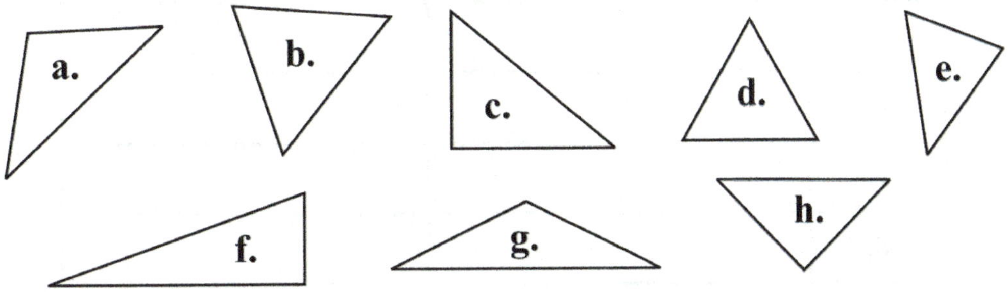

5. Why doesn't the definition for a right triangle ask for two or three of the angles to be *right* angles?

6. Plot the points, and connect them with line segments to form two triangles. Classify the triangles by their angles and sides.

Triangle 1: (0, 0), (4, 0), (0, 4)

____________________ and

____

Triangle 2: (5, 5), (1, 8), (9, 4)

____________________ and

____

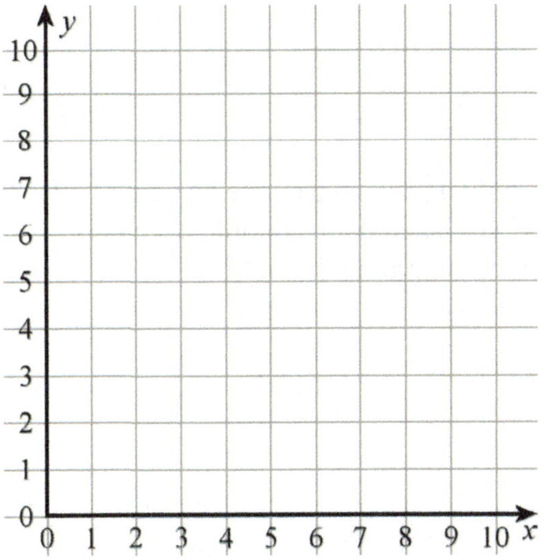

Classifying Triangles 2

1. Classify the triangles as "acute," "right," or "obtuse" (by their angles), and also as "equilateral," "isosceles," or "scalene" (by their sides).

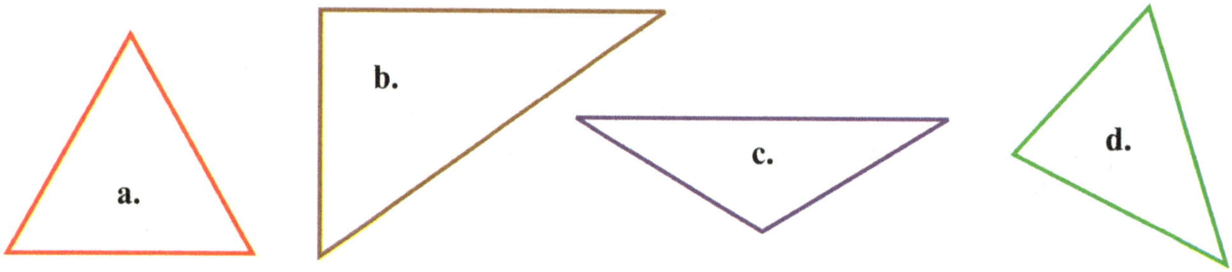

Triangle	Classification by the angles	Classification by the sides
a.		
b.		
c.		
d.		

2. Could an isosceles triangle be acute? Right? Obtuse?
If yes, sketch an example (of each).
If not, explain why not.

3. Could a scalene triangle be acute? Right? Obtuse?
If yes, sketch an example (of each).
If not, explain why not.

4. Plot in the grid:

   a. a right isosceles triangle

   b. an acute isosceles triangle.

5. Fill in the missing parts in this tree diagram classification for triangles.

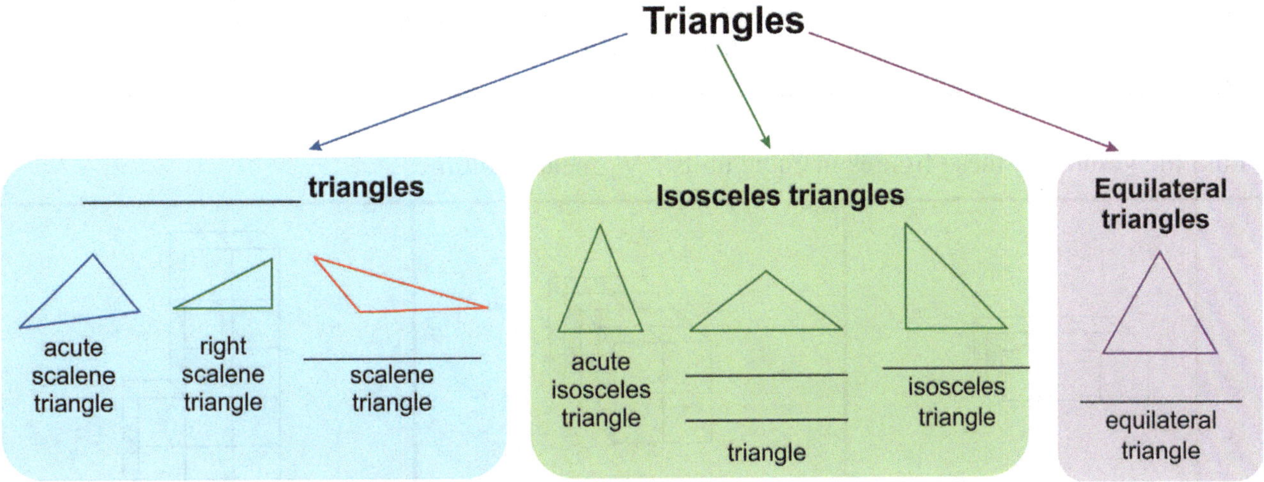

6. An equilateral ("same-sided") triangle is also an equi*angular* ("same-angled") triangle. Make a guess about the angle measures in an equilateral triangle, then measure to check. (You can use, for example, the equilateral triangle on the previous page, and continue its sides.)

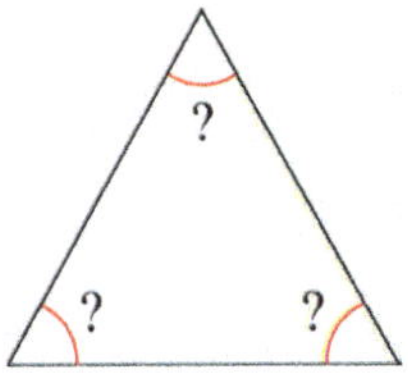

7. When you draw a diagonal into any quadrilateral, you always get two triangles. Sketch a quadrilateral, and a diagonal in it, so that the triangles are acute and scalene.

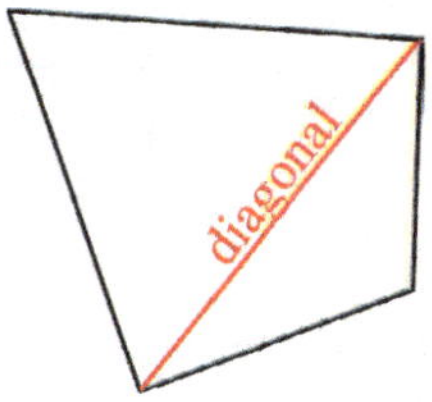

Puzzle Corner

When a certain type of a quadrilateral is divided by a diagonal into two triangles, the triangles will always be isosceles. What kind of quadrilateral is this?

Volume

The **volume** of an object has to do with how much SPACE it takes up or occupies.

You have measured the volume of liquids with measuring cups that use ounces or milliliters. If we need to know the volume of a big object, such as a room, we cannot pour water into it to measure it with measuring cups. Instead, we use cube-shaped units or **cubic units**, and we simply check or calculate how many cubic units fit into the object.

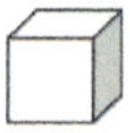

This little cube is **1 cubic unit.**

The volume of the figure on the right is six cubic units. We write V = 6 cubic units. Note that one cube is not visible.

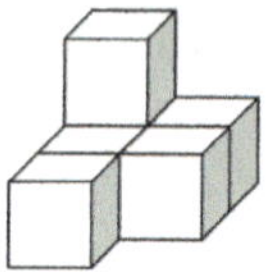

1. Find the volume of these figures in cubic units. “V” means volume.

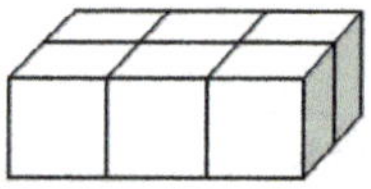

a. V = ________ cubic units

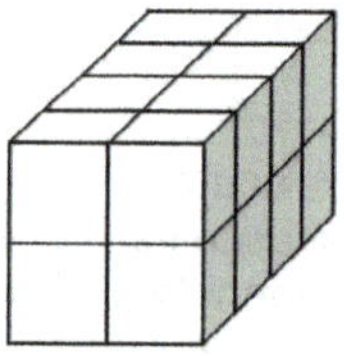

b. V = ________ cubic units

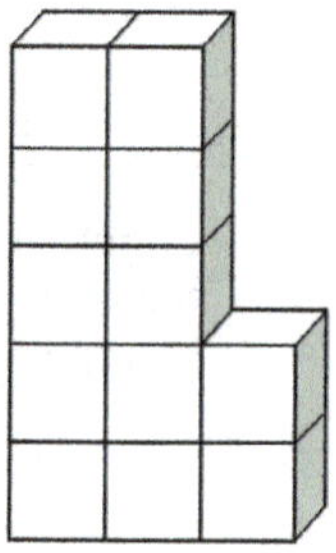

c. V = ________ cubic units

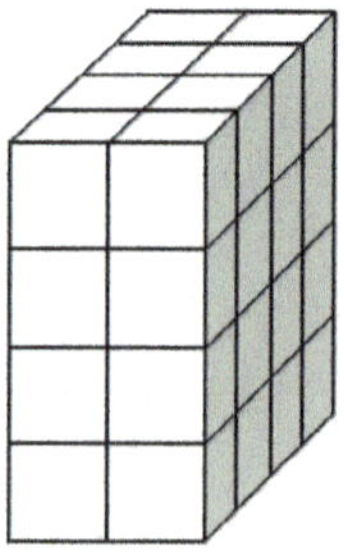

d. V = ________ cubic units

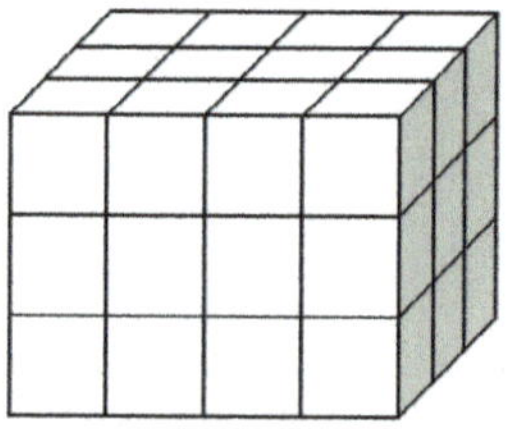

e. V = ________ cubic units

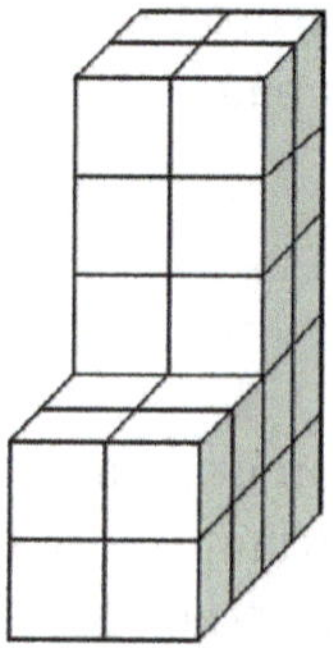

f. V = ________ cubic units

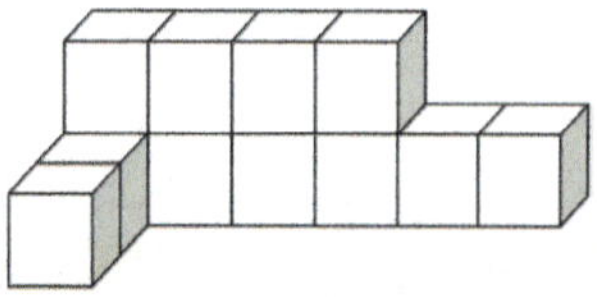

g. V = ________ cubic units

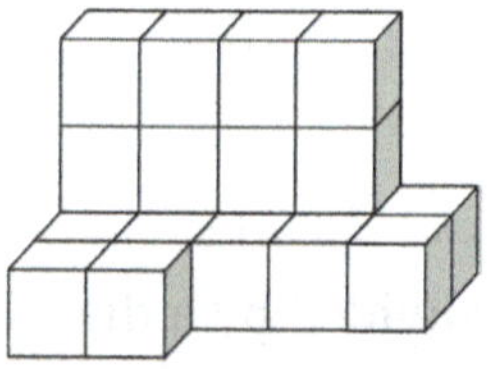

h. V = ________ cubic units

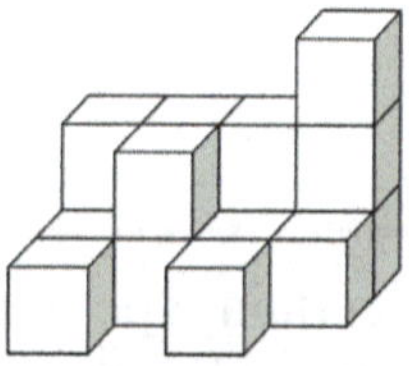

i. V = ________ cubic units

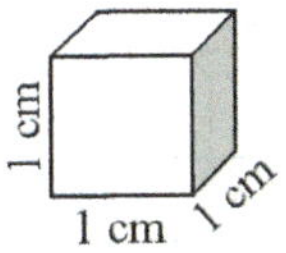

Each edge of this cube measures 1 cm. The volume of the cube is **1 cubic centimeter**. This is abbreviated as **1 cm³**.

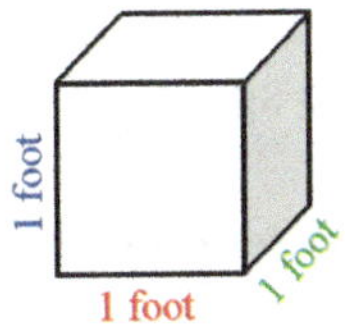

Each edge of this cube is 1 foot. Its volume is **1 cubic foot.**

V = 1 cu. ft. or 1 ft^3

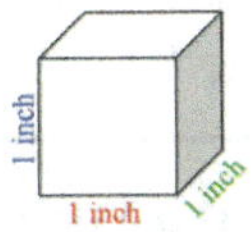

Each edge of this cube is 1 inch. Its volume is **1 cubic inch**.

V = 1 cu. in. or 1 in^3

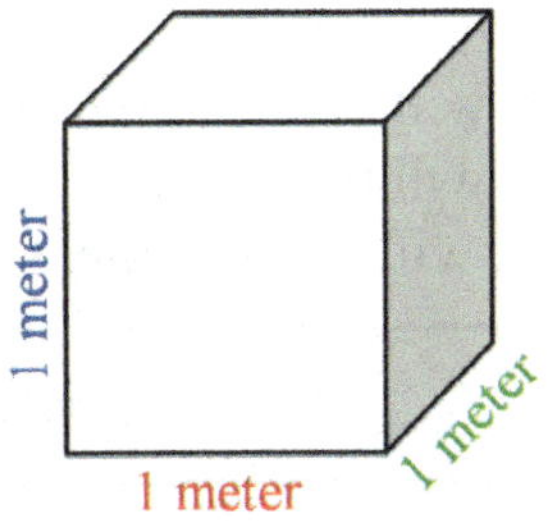

Each edge of this cube is 1 meter. Its volume is **1 cubic meter.**

V = 1 m^3

In general, if the edges of the cube are in certain units (such as inches, feet, centimeters, or meters), then the volume will be in corresponding *cubic* units.

If no unit is given for the edge lengths, we use the word "unit" for the lengths of the edges, and "cubic unit" for the volume. This "box" has a volume of 18 cubic units.

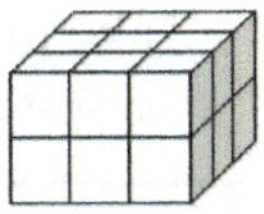

2. Find the total volume of each figure when the edge length of the little cube is given. Remember to include the unit!

The edge of each cube is 1 in.

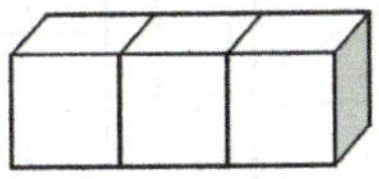

a. V = $3\ in^3$

The edge of each cube is 1 ft.

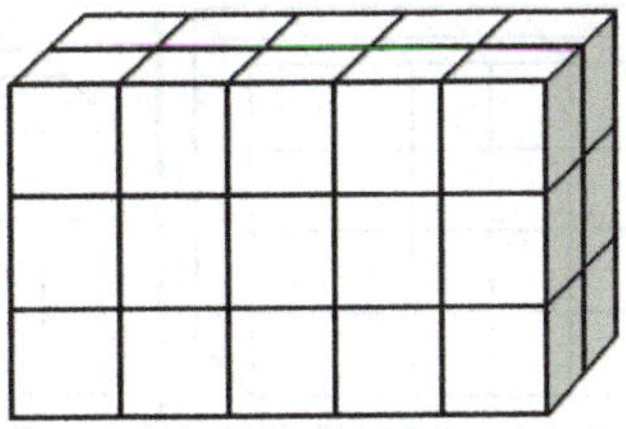

b. V = ______________

The edge of each cube is 1 cm.

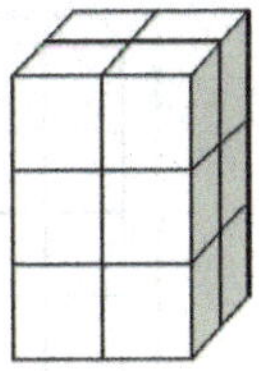

c. V = ______________

The edge of each cube is 1 m.

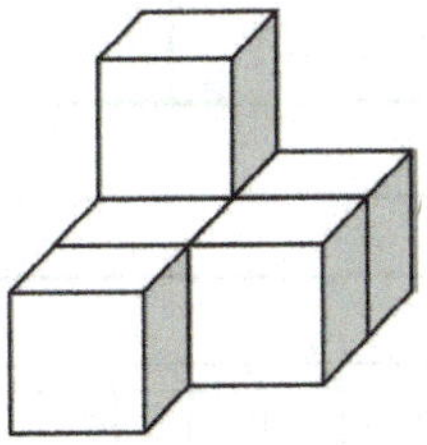

d. V = ______________

The edge of each cube is 1 cm.

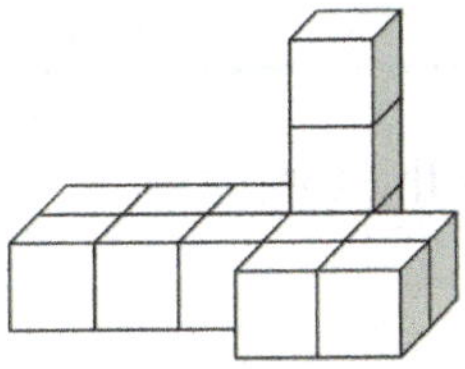

e. V = ______________

The edge of each cube is 1 in.

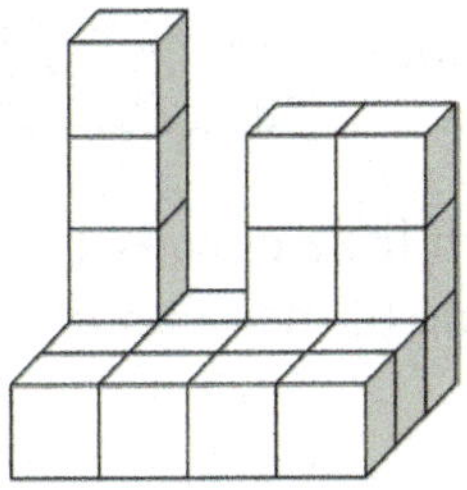

f. V = ______________

This figure is called a **rectangular prism.** It is also called *a cuboid.*
It is simply a box with edges that meet at right angles.

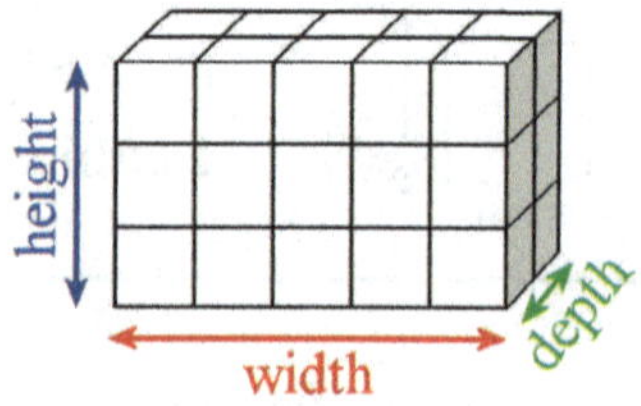

Many people call the **three dimensions** that we measure "length," "width," and "height." Here we will use "width," "depth," and "height."

The **width** will be the dimension that runs left to right.
The **depth** will be the dimension that points away from you—into the paper, so to speak.
The **height** will be the dimension pointing "up" in the figure.

A way to find the volume of a rectangular prism

1) Can you figure out a way to find the number of cubes in the *bottom* layer of this rectangular prism *without* counting?

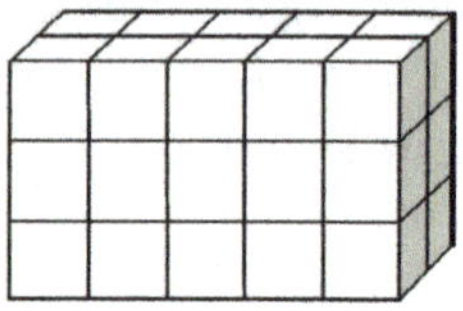

You can multiply 5 × 2 = 10, which means multiplying the *width* and the *depth*. The bottom layer has 10 cubic units.

2) After that, there is a way to easily find the *total* number of cubes in the rectangular prism (its volume). Can you figure that out?

Take the number of cubes in the bottom layer, and **multiply that by the number of layers** (the *height*). There are 10 cubes in the bottom layer, and 3 layers. We get 10 × 3 = 30 cubic units.

3. Find the volume of these rectangular prisms by finding the amount of cubic units in the bottom layer and multiplying that by the height (how many layers there are).

	a.	**b.**	**c.**	**d.**
	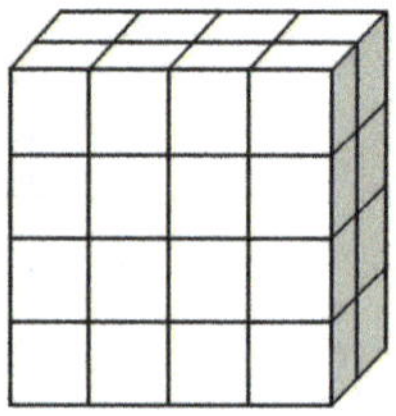	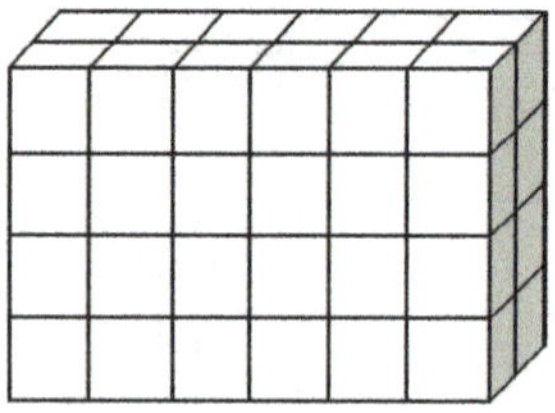	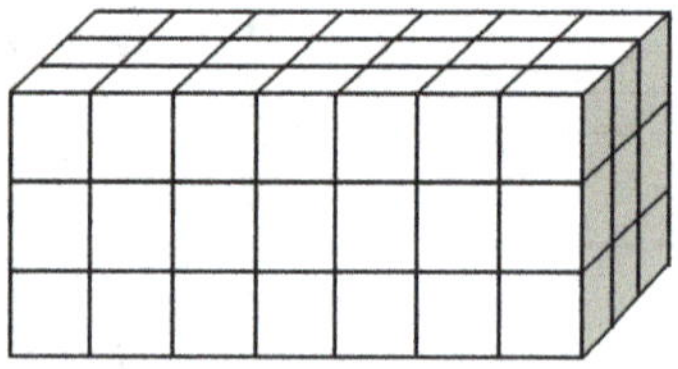	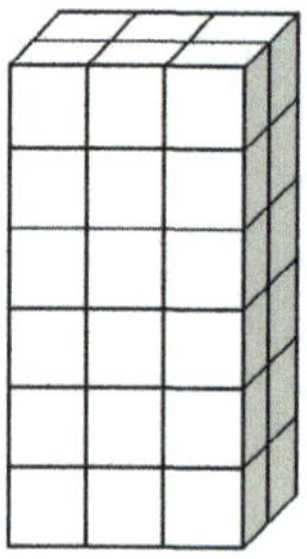
Cubes in the bottom layer	*8*			
Height	*4*			
Volume	*32*			

4. If each little cube is 1 cubic inch, what is the total volume of the outer box?

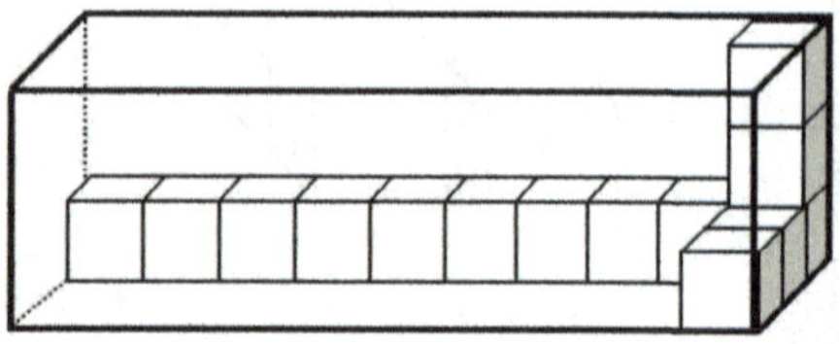

Notice what we did in these two steps:

(1) We multiplied the width and the depth to find the number of cubes in the bottom layer. **Multiplying the width and the depth** also gives us **the area of the bottom** (A_b)! For example, the bottom area of this cuboid is $4 \times 3 = 12$ square units.

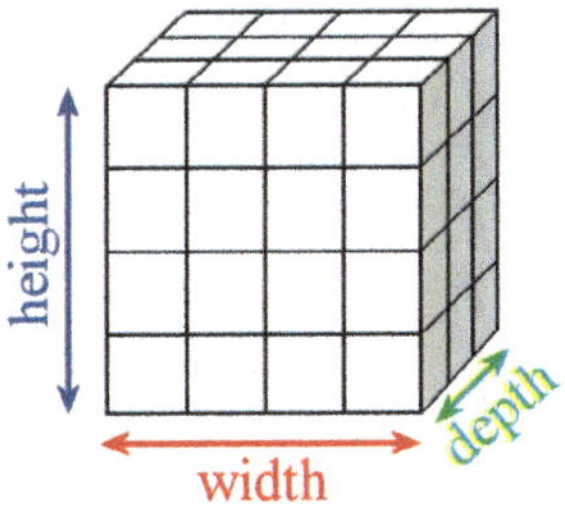

(2) We multiplied what we got from step 1 by height.

We ended up multiplying the bottom area by the height.
Or, looking at it in another way, we multiplied the width, the depth, and the height.

From that we get two **formulas** for the volume of a rectangular prism:

1. $V = w \times d \times h$ (volume is width × depth × height)
2. $V = A_b \times h$ (volume is area of the bottom × height)

5. Write the width, height, depth, and volume of these rectangular prisms.

	a. 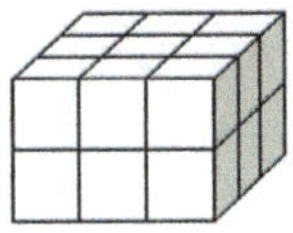	**b.** 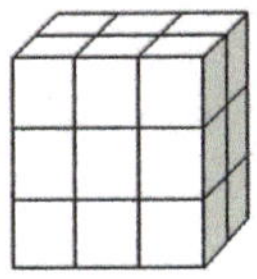	**c.** 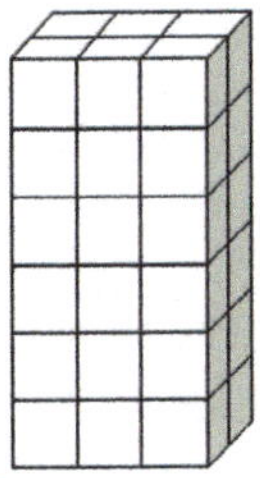	**d.** 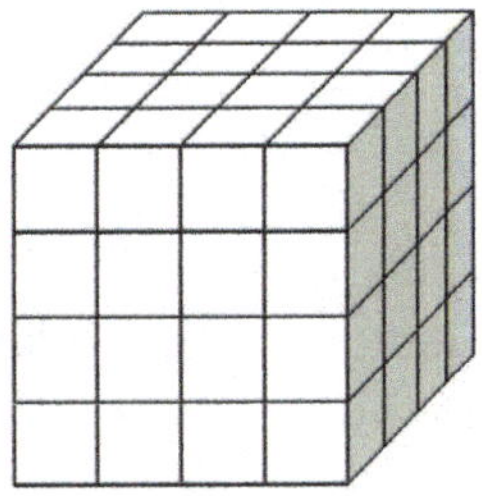
Width:				
Depth:				
Height:				
Volume:				

6. Find the volume of the rectangular prisms above *if* their top layer was removed.

a. V = _______ cubic units **b.** V = _______ cubic units

c. V = _______ cubic units **d.** V = _______ cubic units

7. Erica has started filling a box with little cubes that are one cubic centimeter each. The outer box is 5 cm high. How many more little cubes does she need to fill the box?

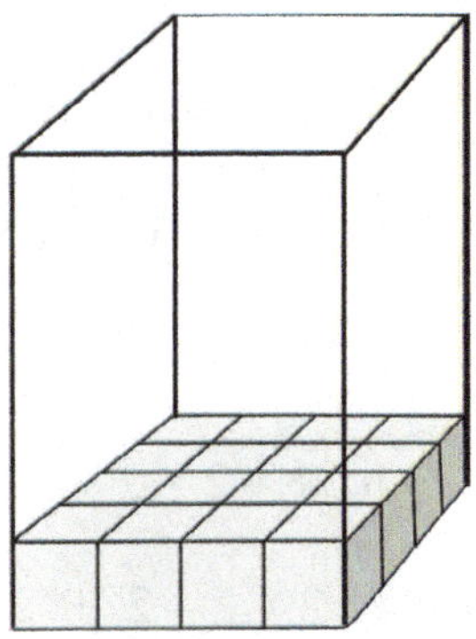

Now we can explain where the little raised "3" (the exponent) in cubic units comes from.

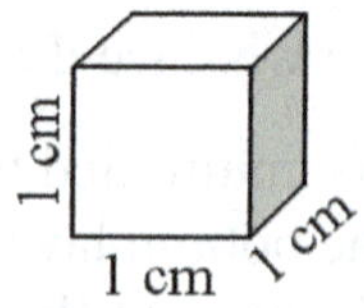

To find the volume of this cube, we multiply its width, the height, and the depth. This means multiplying 1 cm × 1 cm × 1 cm. Not only do we multiply number 1 by itself three times—we also multiply the *unit centimeter* (cm) by itself three times. The little "3" in cm^3 shows that.

8. **a.** Sketch a rectangular prism with a volume of 4 × 2 × 6 cubic units.

   b. Sketch a rectangular prism with a volume of 3 × 3 × 3 cubic units.

   c. Sketch a rectangular prism with a volume of 2 × 5 × 4 cubic units.

9. Chris and Mia drew these rectangular prisms to match the expression 5 × 3 × 2. Who is right?

Chris:

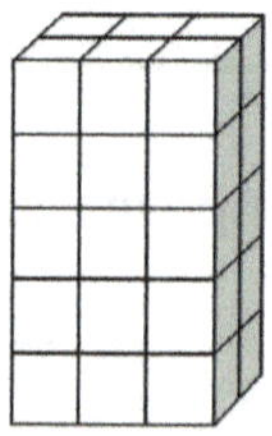

Mia:

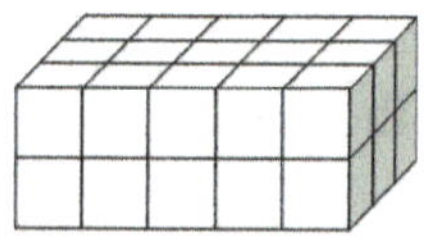

10. To calculate the volume of this kind of figure, think of it as consisting of *two* rectangular prisms. We calculate the volume of each separately, and then add. Which expression below matches the volume of this figure?

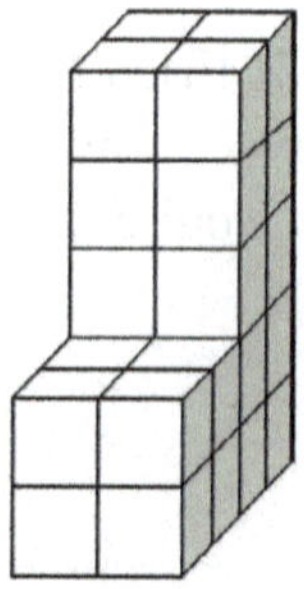

   a. 2 × 3 × 2 + 2 × 2 × 3

   b. 2 × 2 × 2 + 2 × 2 × 3

   c. 2 × 2 × 2 + 2 × 2 × 5

Volume of Rectangular Prisms

Study the two formulas for the volume of a rectangular prism:	1. $\mathbf{V = w \times d \times h}$	(volume is width × depth × height) *Some people use width, length, and height instead.*
	2. $\mathbf{V = A_b \times h}$	(volume is area of the bottom × height)

The width, depth, and height need to be in the same kind of unit of length (such as meters). The volume will then be in corresponding cubic units (such as cubic meters).

Example 1. A room measures 12 ft by 8 ft, and it is 8 ft high. What is the volume of the room? What is the area of the room?

To find the area, we simply multiply the two given dimensions: A = 12 ft × 8 ft = 96 ft^2.
To find the volume, we can multiply the area by the height: V = 96 ft^2 × 8 ft = 768 ft^3.

1. **a.** Find the volume of a box that is 2 inches high, 5 inches wide, and 7 inches deep. Include the units! V = *5 in* × ________ × ________ = ________

   b. Find the area and volume of a room that is 25 ft × 20 ft, and 9 feet high. Include the units!

   A = ________ × ________ = ________

   V = ________ × ________ × ________ = ________

2. Find the volume of the boxes with the given dimensions. (*Remember that all the dimensions need to be in the same measurement unit before calculating the volume.*)

   a. 20 cm wide, 30 cm deep, and 0.6 meters high

   b. 16 square inches on the bottom, and half a foot tall

3. The volume of this box is 30 cm^3.
   What is its depth?

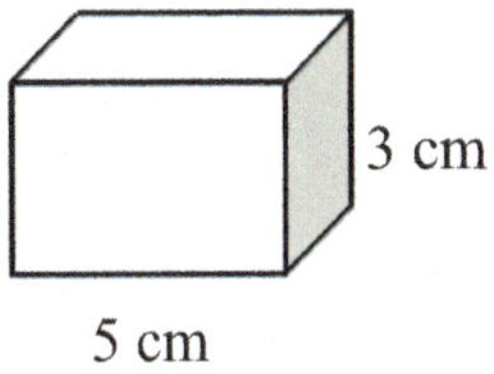

4. *Optional*. Measure the width, height, and depth of a dresser and/or a fridge.
   Find out its volume (in cubic feet or cubic meters).

How to sketch rectangular prisms		
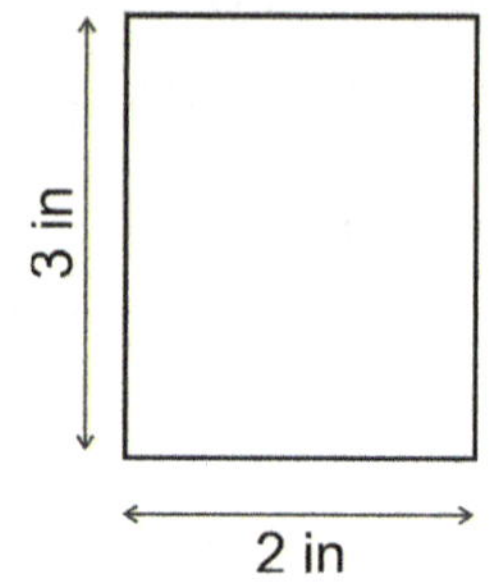 1. Draw a rectangle that has the width and the height of the rectangular prism.	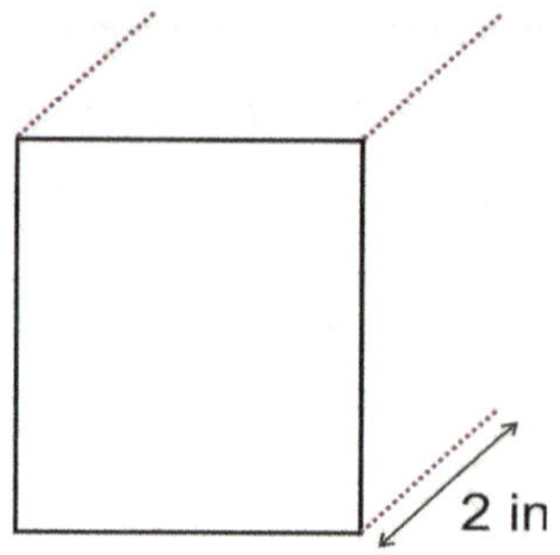 2. Draw lines at about 45° angles to show the depth. Because of the perspective, draw these lines somewhat *shorter* than your ruler would indicate.	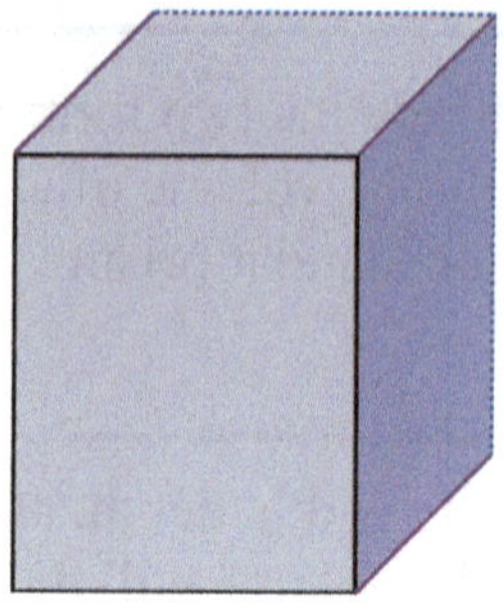 3. Draw the last lines. Shade the faces of your shape if you like.

5. Sketch a box that is 6 cm wide, 5 cm tall, and 4 cm deep. What is its volume?

6. Sketch a box that is 50 cm wide, 80 cm tall, and 90 cm deep. What is its volume?

7. The length and width of a rectangular box are 5 inches and 6 inches. Its volume is 180 cubic inches. How tall is it?

8. The picture shows an aquarium that is 1/4 filled with water. Find the volume of the water in it.

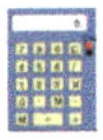

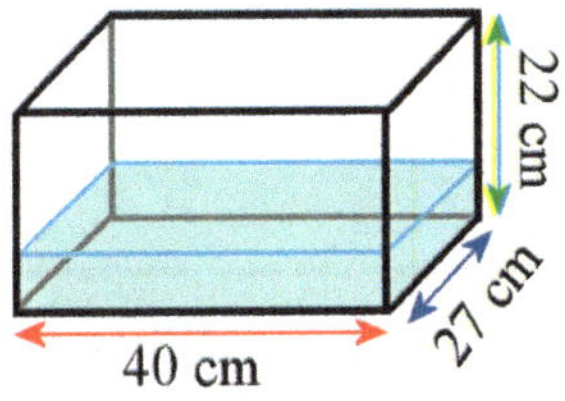

9. **a.** Design a box (give its width, height, and depth) with a volume of 64 cubic inches.

b. Design a water tank, in the form of a rectangular prism, that can hold 960,000 cm^3 of water (which equals 960 liters).

10. Amber built this prism with little cubes. Then she built another, with a volume that was four times the volume of the little prism.

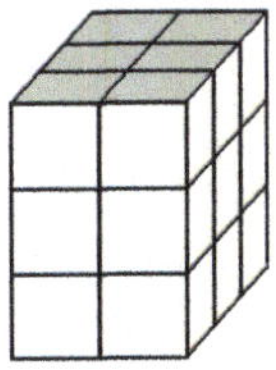

a. What is the volume of the larger prism?

b. What could its dimensions be?

11. John's room is 12 ft × 18 ft, and it is 9 ft high. The family plans to *lower* the ceiling by 1 foot.

a. What will the volume of the room be after that?

b. How much volume will the room lose?

Puzzle Corner

A truck delivered two cubic yards of gravel to Tom. Calculate how many **cubic feet** of gravel Tom got.

Hint: Design a box with a volume of two cubic yards.

Volume Is Additive

Volume is **additive**. What we mean by that is that we can ADD to find the total volume of a shape that is in several parts.

To find the total volume of the shape on the right, first find the volume of the top box, then the volume of the bottom box, and add the two volumes.

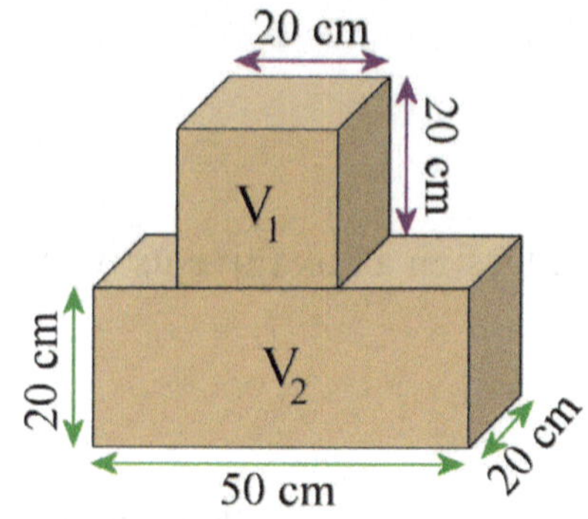

1. Find the total volume of the shape in the teaching box above.
   Show your work, and organize your work carefully, to avoid mistakes.

   $V_1 =$

   $V_2 =$

   $V_{total} = V_1 + V_2 =$

2. This is a two-part kitchen cabinet. Its height is 2 ft and depth 1 ft. One part is 5 ft long, and the other is 4 ft long.

   a. Mark the given dimensions in the picture.

   b. Calculate the volume.

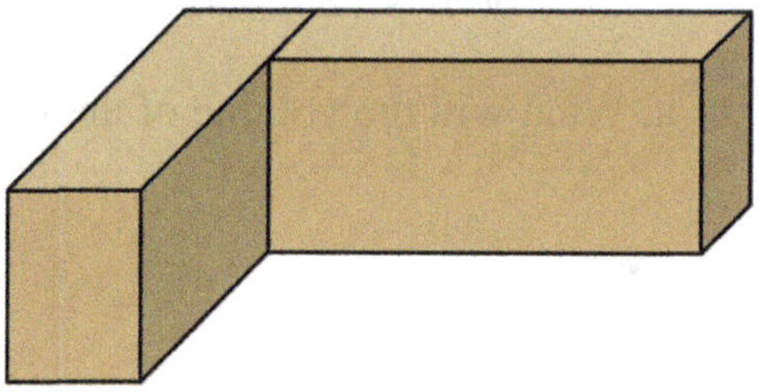

3. Find the volume of this building.

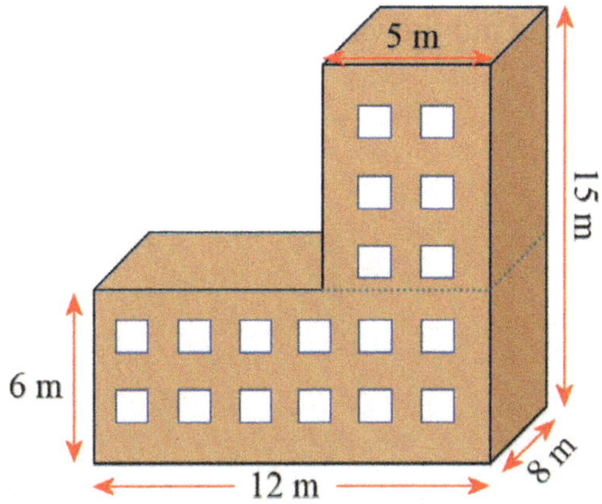

4. Find the total volume.

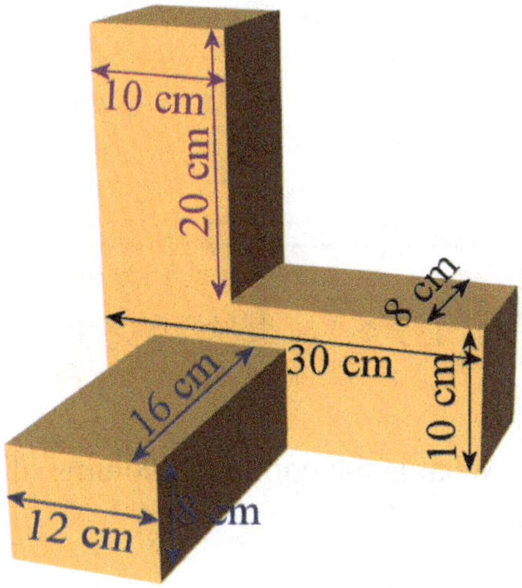

5. **a.** Find the volume of this two-part bird cage.

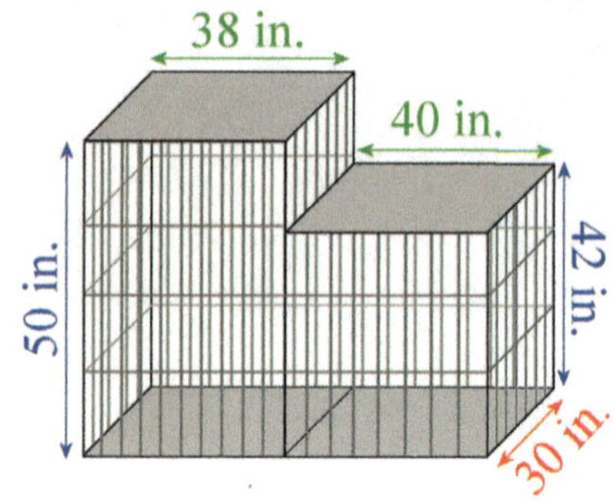

b. One cubic foot is 1,728 cubic inches.
Convert your answer from (a) into cubic feet, to three decimals.

Puzzle Corner

The volume of the larger cube is 1,000 cubic inches. The edge length of the smaller cube is half of the edge length of the larger cube.

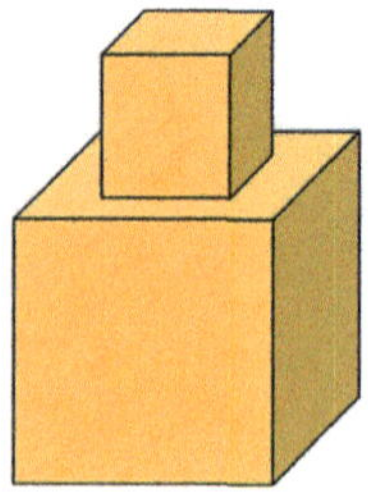

What is the combined volume of the two cubes?

Area and Perimeter Problems

(This lesson is optional.)

Example 1. Find the area of the shaded figure.

The easiest way to do this is:
(1) Find the area of the larger outer rectangle.
(2) Find the area of the white inner rectangle.
(3) Subtract the two.

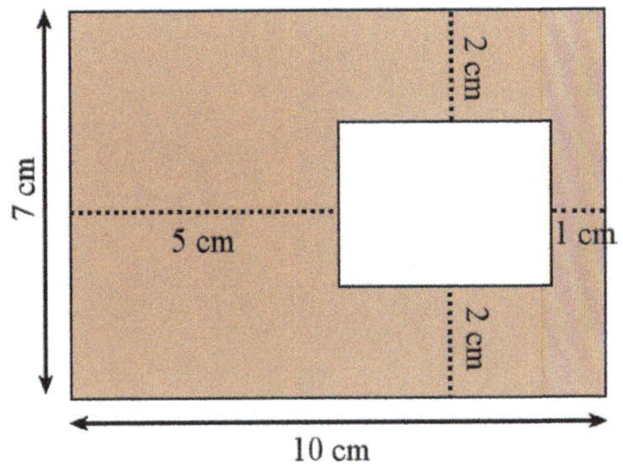

1. The area of the large rectangle is
   7 cm × 10 cm = 70 cm^2.

2. First we find the *sides* of the white rectangle
   by subtracting.
   The longer side is 10 cm − 5 cm − 1 cm = 4 cm.
   The shorter side is 7 cm − 2 cm − 2 cm = 3 cm.

   So, the area of the white rectangle is 4 cm × 3 cm = 12 cm^2.

3. Lastly we subtract the two areas to find the shaded area: 70 cm^2 − 12 cm^2 = 58 cm^2.

Note: In the following problems, record your work—all the calculations—carefully. That way you are much less likely to make mistakes! Use a notebook, if necessary, for additional space.

1. **a.** Find the area of the *white* rectangle.
   All lines meet at right angles.

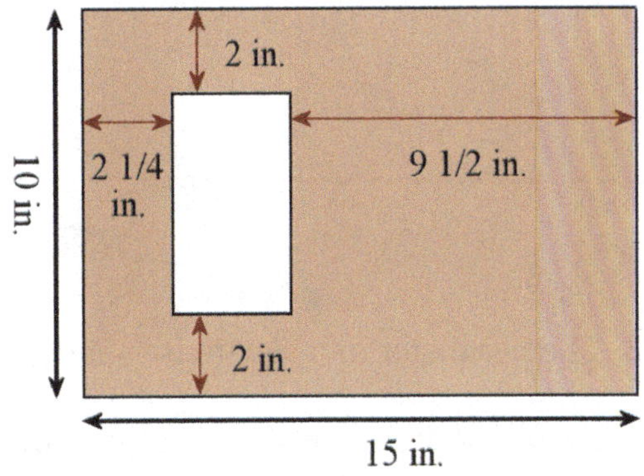

   b. Find the area of the shaded figure.

2. The image on the right shows a picture frame.
   Find the area of the actual frame (that is, of the shaded part).
   Give your answer to the nearest whole square centimeter.
   (All lines meet at right angles.)

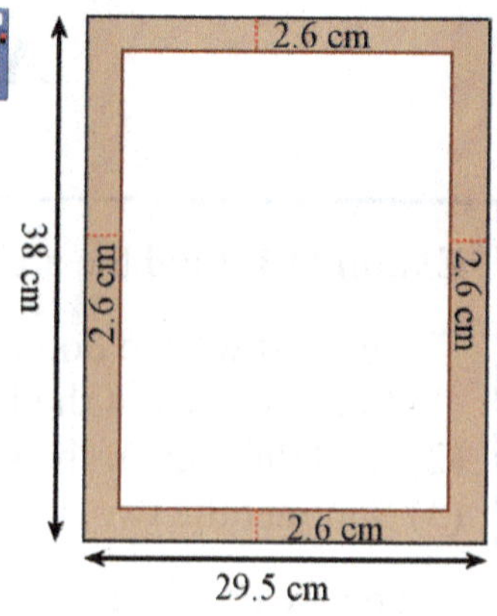

3. The perimeter of a square is ½ mile.

   a. How long is one side of the square, in miles?
   Draw a sketch to help you.

   b. How long is one side of the square, in *feet*?
   (1 mile = 5,280 ft)

4. A wall is 16 feet wide and 10 feet high. It has one 3.5 ft × 4.5 ft window in it.

a. Draw a sketch.

b. Find the area of the window.

c. What is the area of the actual wall (not including the window)?

d. A gallon of paint covers 350 square feet of wall.
How many whole gallons do you need to paint the wall?

e. If instead of gallons, you buy paint by quarts, how many whole quarts of paint do you need to paint the wall?

Example 2. Find the perimeter of the figure.

We need to find the length of *each* side and then add the lengths. Start, for example, at the side marked with 1, then go to the side marked with 2, then to side 3, and so on, until you have "traveled" all the way around the figure.

Side 1 is 3 cm. Side 2 is 2 cm. Side 3 is 5 cm.
The total perimeter is:

3 cm + 2 cm + 5 cm + 5 cm + 4 cm + 1 cm + 4 cm + 4 cm = 28 cm.

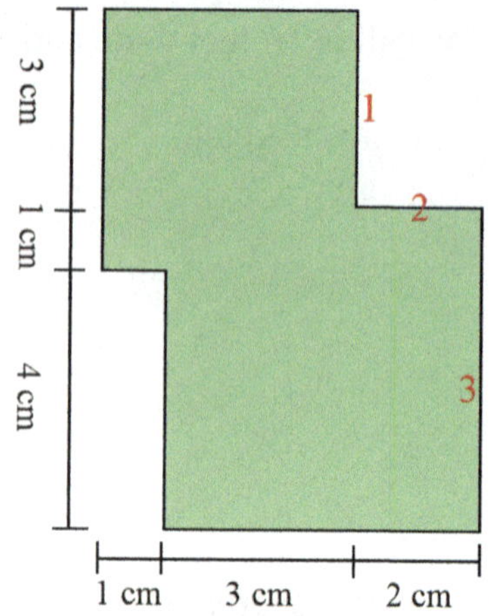

5. Find the area of the figure. The figure has already been divided into rectangles to help you.

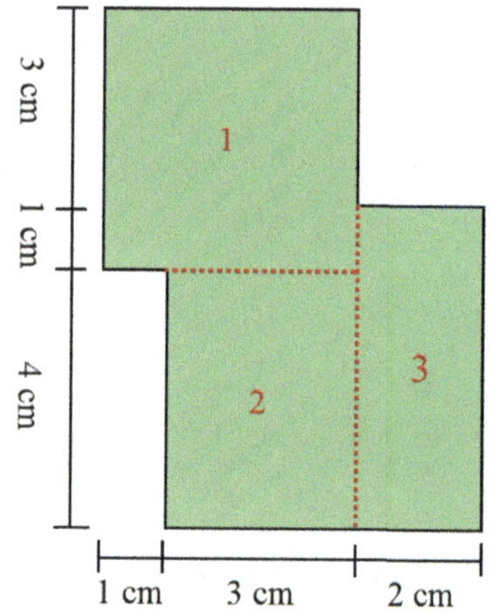

6. Find the area and the perimeter of this figure. All lines meet at right angles.

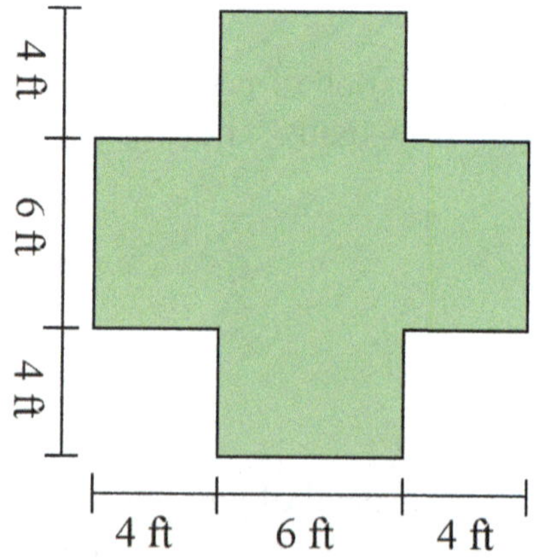

7. Find the area and the perimeter of this figure.
All lines meet at right angles.
The dimensions are given in centimeters.

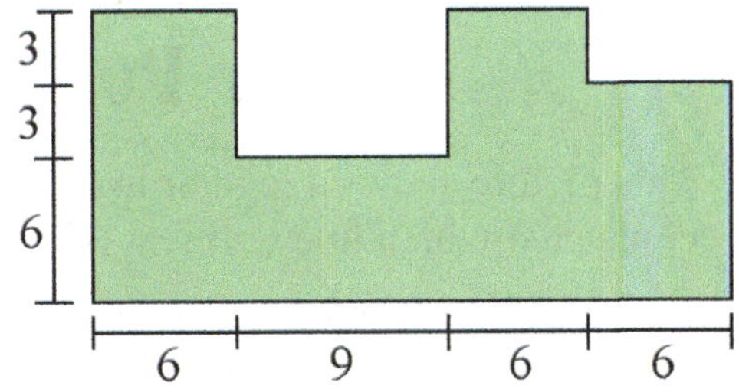

8. Find the area and the perimeter of this figure.
All lines meet at right angles.
The dimensions are given in inches.

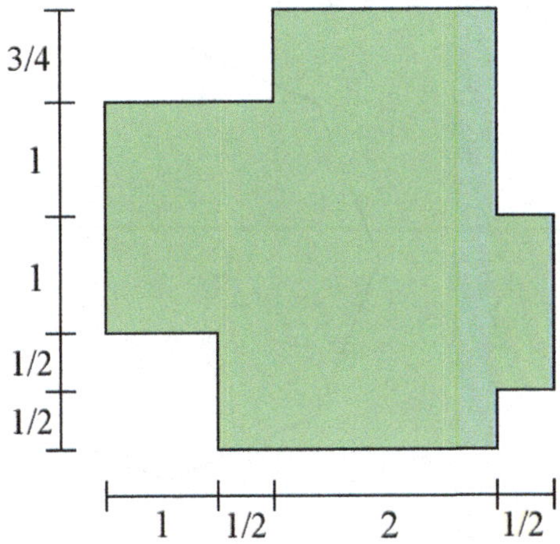

Star Polygons (*optional*)

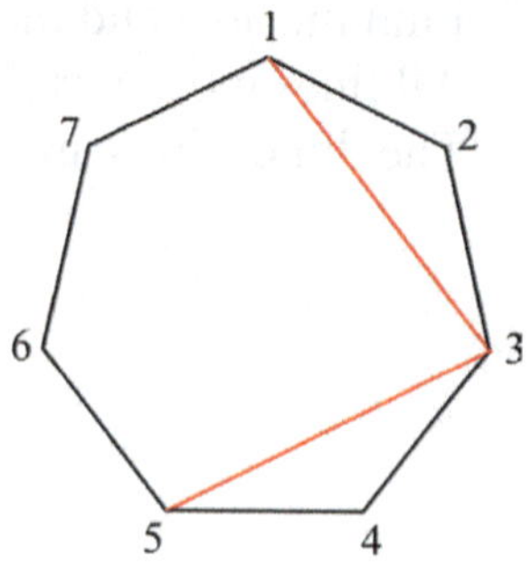

1. This picture shows a regular heptagon where every *other* vertex is connected with a line segment (skipping one vertex in between).

   Continue drawing diagonals in such a manner. The shape you will get is called a *star polygon*, and specifically a *heptagram*.

2. Make star polygons.

   a. This is a regular nonagon. Make a nonagram connecting every four vertices (skipping 3).

   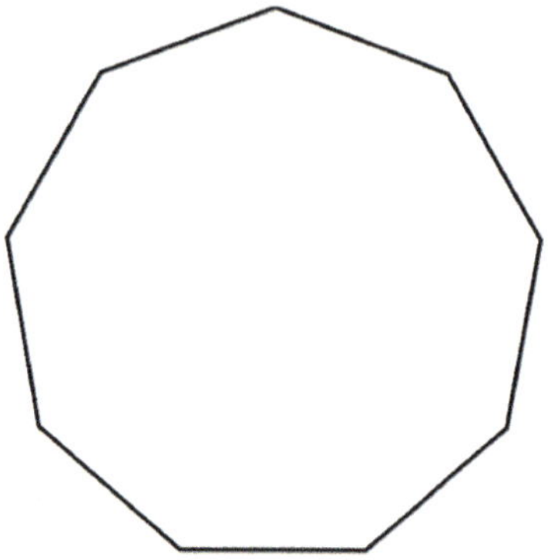

   b. This is a regular pentagon. Make a pentagram connecting every other vertex (skipping one).

   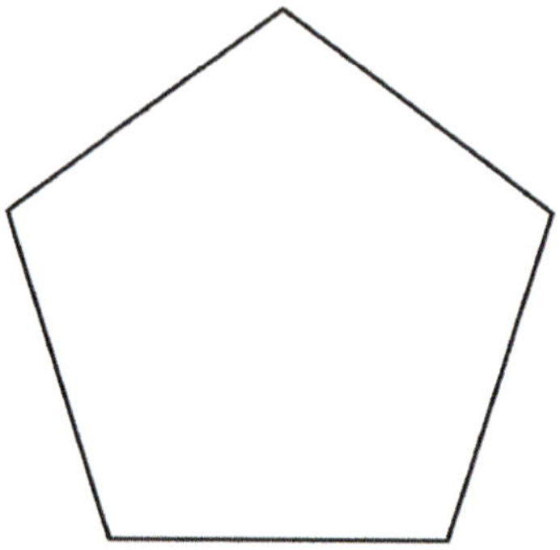

   c. This is a regular octagon. Make an octagram connecting every three vertices (skipping 2).

   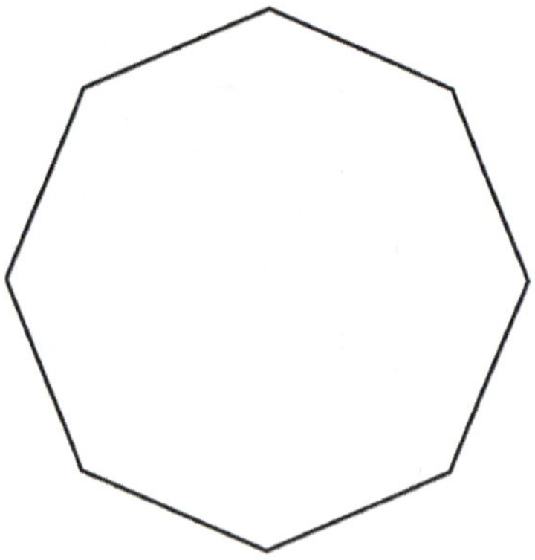

   d. This is a regular heptagon. Make a heptagram connecting every three vertices (skipping 2).

   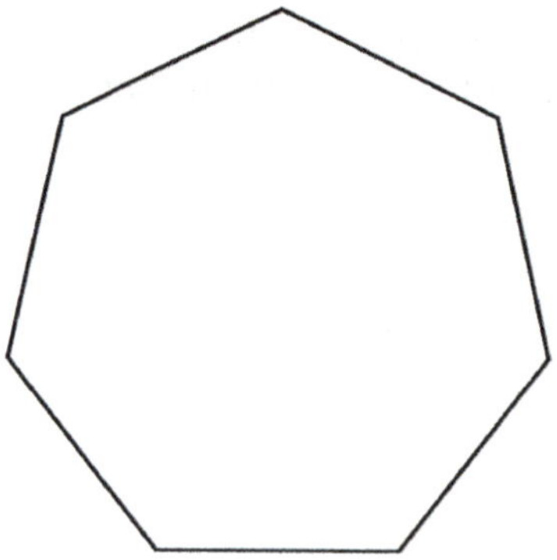

   e. This is a regular decagon. Make a decagram connecting every three vertices (skipping 2).

   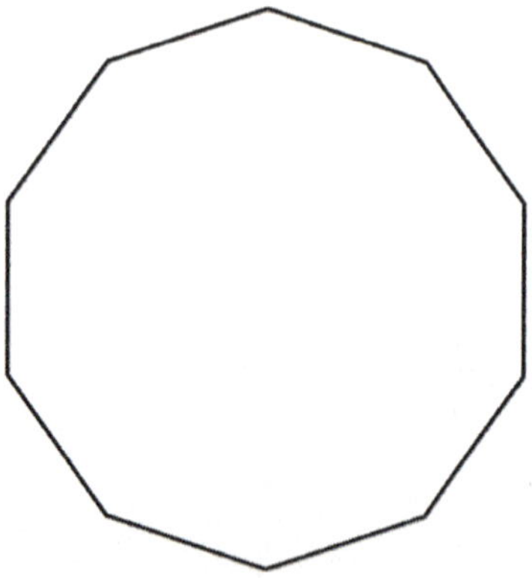

   f. Make a nonagram connecting every second vertex (skipping 1).

   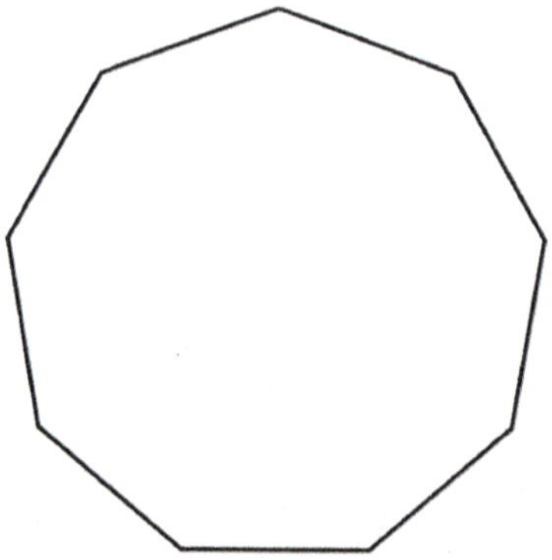

3. You can make your own star polygons here. Experiment!

Review 1

1. A farmer has 45 feet of fencing to fence in a chicken yard. He is only using the fencing on three sides of the yard, and the fourth side will be the wall of the chicken house (10 ft).

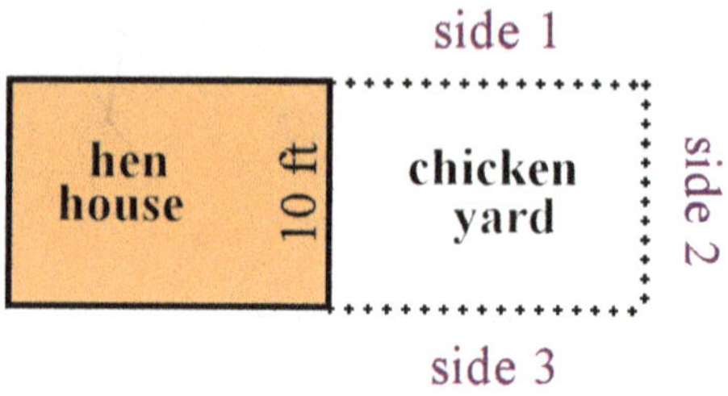

   How long are the sides of the yard?

   Side 1:

   Side 2:

   Side 3:

2. Find the area of the shape.

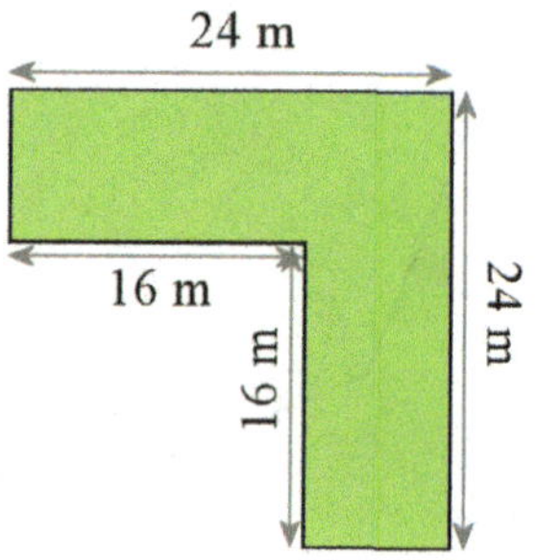

3. Measure the angles.

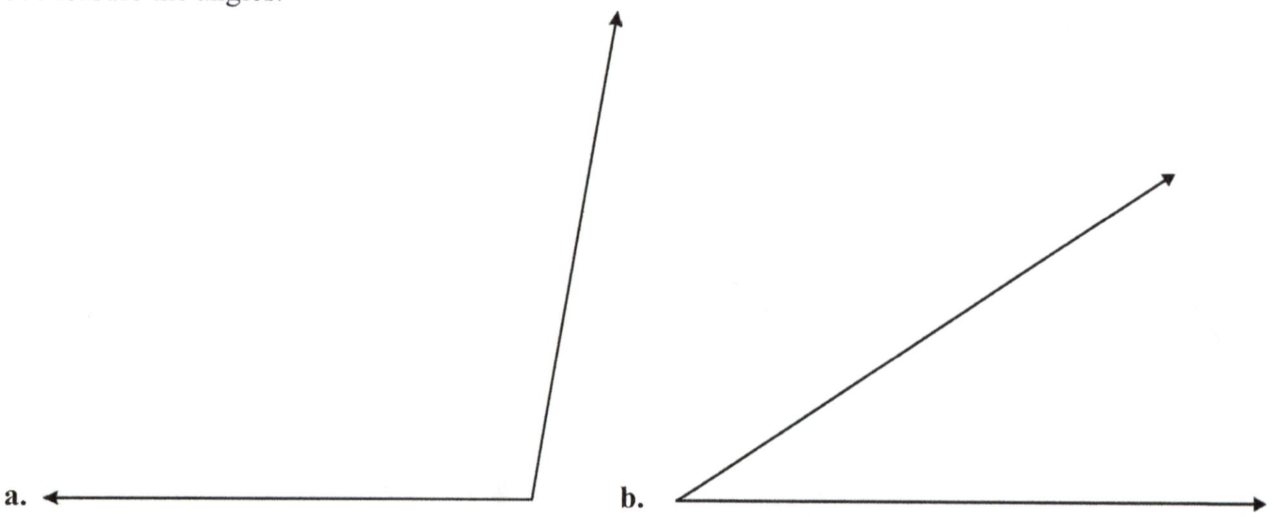

4. Label each angle as acute, obtuse, or right.

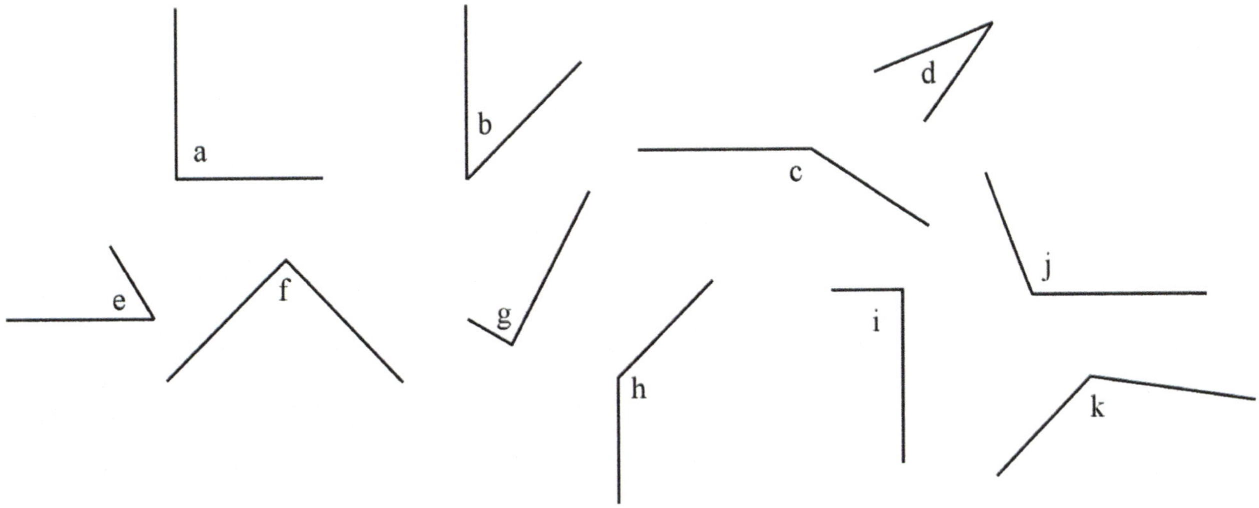

5. Draw an angle that measures 65°.

6. Figure out the unknown angle measure.

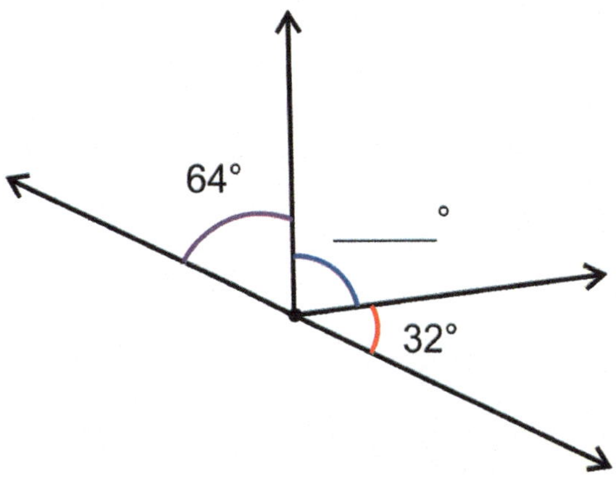

7. Draw a line that is perpendicular to this line and goes through the given point.

8. Find the task that is possible to do, and complete it.

   a. Draw a rectangle with one 110° angle.

   b. Draw a triangle with one 110° angle.

   c. Draw an acute angle that measures 110°.

9. **a.** Draw any obtuse triangle.

   b. Measure all the angles of your triangle.

10. Draw ONE **diagonal** (a line from corner to corner) into this square.
You will get two triangles.
What kind of triangles are they?

11. Find rays, lines, and line segments that are either parallel or perpendicular to each other. Use ‖ for parallel and ⊥ for perpendicular.

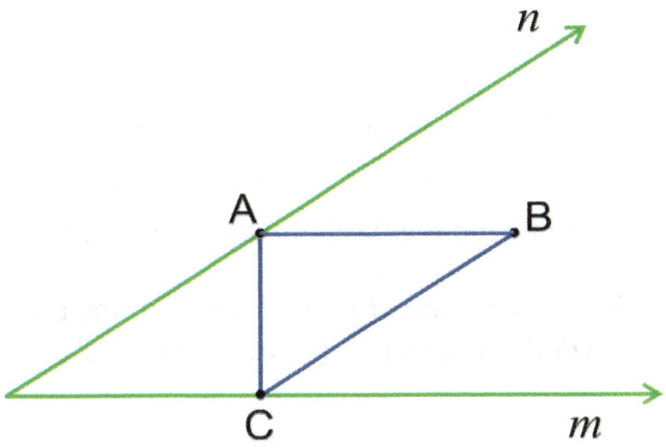

12. Draw any parallelogram.

Hint: First, draw a line. Then draw a line that is parallel to that line. Then draw a third line that intersects the other two.

13. Are these figures symmetrical? Draw a symmetry line or lines in the ones that are.

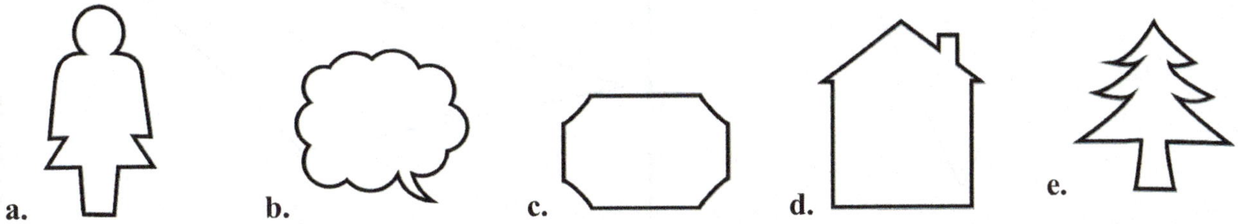

Review 2

1. What is the name of a polygon...

   a. with five sides?

   b. with eight sides that are all the same length?

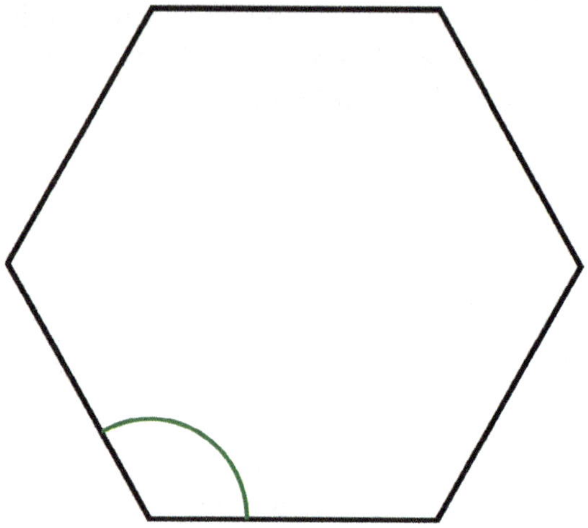

2. **a.** Name the regular polygon on the right.

   b. Measure one of its angles to find the measure of *all* the angles in it (since it's regular!).

3. Name the shape below, and find out the measures of its angles.

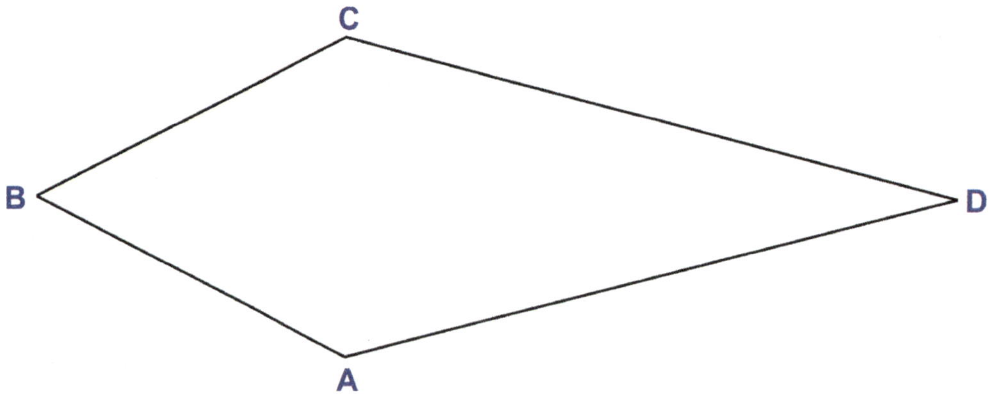

Angle A = ________° Angle B = ________° Angle C = ________° Angle D = ________°

4. Classify each triangle according to its sides and its angles.

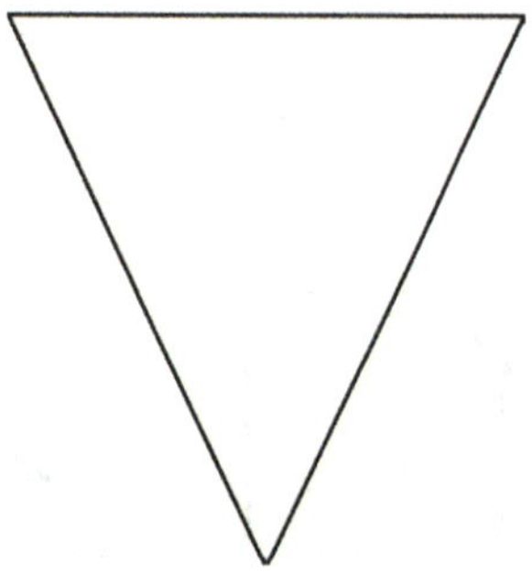

a. ______________________________

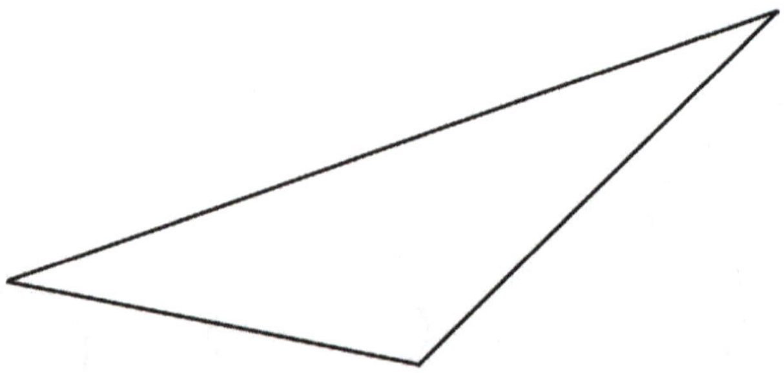

b. ______________________________

5. Could an equilateral triangle be acute? Right? Obtuse?

6. **a.** Sketch an isosceles obtuse triangle.

b. Sketch a scalene acute triangle.

7. Name the different types of quadrilaterals.

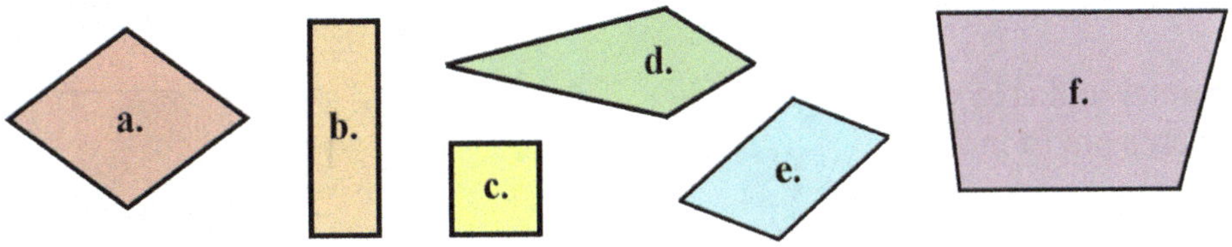

a. | **b.**

c. | **d.**

e. | **f.**

8. Name the quadrilateral that...

a. is a parallelogram and has four right angles.

b. is a parallelogram and has four congruent sides.

c. has two parallel sides and two sides that are not parallel.

9. Find the volume of this rectangular prism, if…

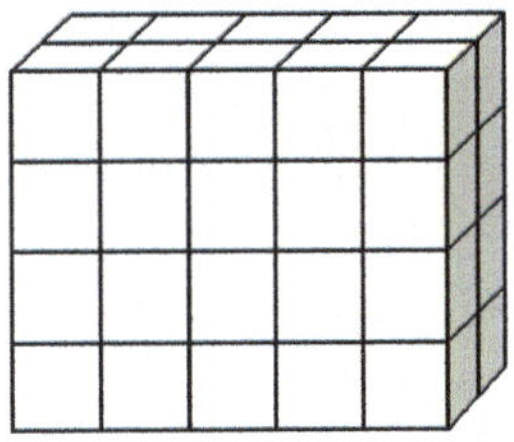

a. …the edge of each little cube is 1 inch.

b. …the edge of each little cube is 2 inches.

10. What is the height of this box, if its bottom dimensions are 2 cm × 4 cm and its volume is 32 cubic centimeters?

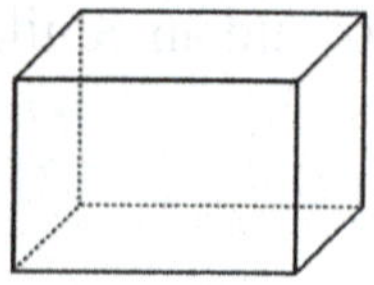

11. A gift box is 6 inches wide, 3 inches deep, and 2 inches tall. How many of these boxes do you need to have a total volume of 108 cubic inches?

12. Find the perimeter and area of this figure. All measurements are in inches.

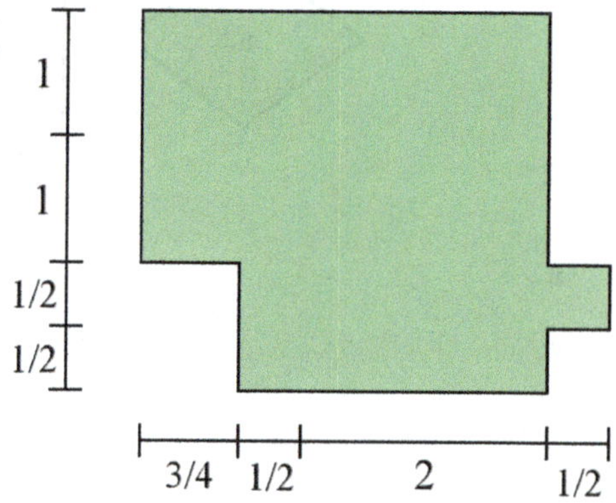

13. **a.** Draw a triangle with 90° and 45° angles.

b. Measure the third angle. It measures ________°.

c. Label each angle in the triangle as acute, obtuse, or right.

Puzzle Corner The area of the bottom face of a cube is 16 cm^2. What is its volume?

Answers

Review: Area of Rectangles, pp. 8-10

1. a. $3 \text{ cm} \times 2 \text{ cm} = 6 \text{ cm}^2$ b. $1 \text{ ft} \times 1 \text{ ft} = 1 \text{ ft}^2$ c. $4 \text{ km} \times 8 \text{ km} = 32 \text{ km}^2$

2. a. 5 in b. 10 cm c. 25 ft

3.

a. $12 \text{ cm} \times s = 96 \text{ cm}^2$ $s = 8$ cm	b. $20 \text{ cm} \times s = 1{,}000 \text{ cm}^2$ $s = 50$ cm

4. a. $9 \text{ m} \times s = 45 \text{ m}^2$ $s = 5$ m b. $60 \text{ ft} \times s = 1{,}800 \text{ ft}^2$ $s = 30$ ft

5. a. $36 \text{ in} \times 24 \text{ in} = 864 \text{ in}^2$ (the student needs to use the multiplication algorithm with pencil and paper)
   b. $3 \text{ ft} \times 2 \text{ ft} = 6 \text{ ft}^2$

6. a. $4 \times (2 + 5) = 4 \times 2 + 4 \times 5 = 8 + 20 = 28$ b. $3 \times (5 + 2) = 3 \times 5 + 3 \times 2 = 15 + 6 = 21$

7. a. $A = 30 \text{ ft} \times (30 \text{ ft} + 75 \text{ ft})$
   $= 30 \text{ ft} \times 30 \text{ ft} + 30 \text{ ft} \times 75 \text{ ft}$
   $= 900 \text{ ft}^2 + 2{,}250 \text{ ft}^2 = 3{,}150 \text{ ft}^2$
   b. $2{,}250 \text{ ft}^2 - 900 \text{ ft}^2 = 1350 \text{ ft}^2$

8. Check the student's drawings. a. 2 in. b. 3 cm c. 1 ft

Problem Solving: Area of Rectangles, pp. 11-12

1. a. There are several possible number sentences, depending on how you divide the shape into rectangles.
   $4 \times 5 + 3 \times 2 + 2 \times 6 = 20 + 6 + 12 = 38$ square units or $4 \times 2 + 8 \times 3 + 2 \times 3 = 8 + 24 + 6 = 38$ square units
   b. Here it is handy to use subtraction. $7 \times 6 - 3 \times 2 = 36$ square units

2. a.

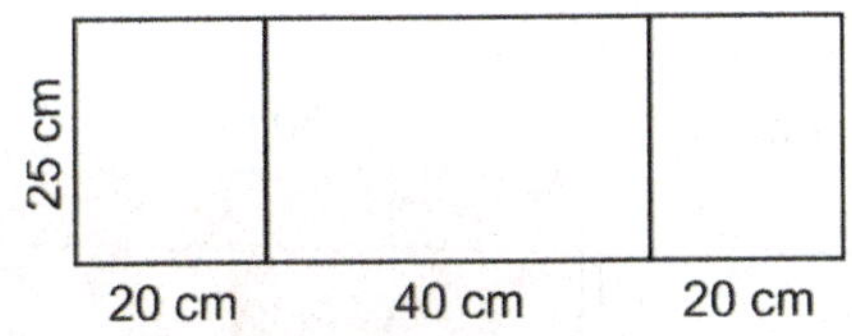

   b. $25 \text{ cm} \times (20 \text{ cm} + 40 \text{ cm} + 20 \text{ cm}) = 25 \text{ cm} \times 80 \text{ cm} = 2{,}000 \text{ cm}^2$.

3. Answers will vary. Check the student's answers. For example:

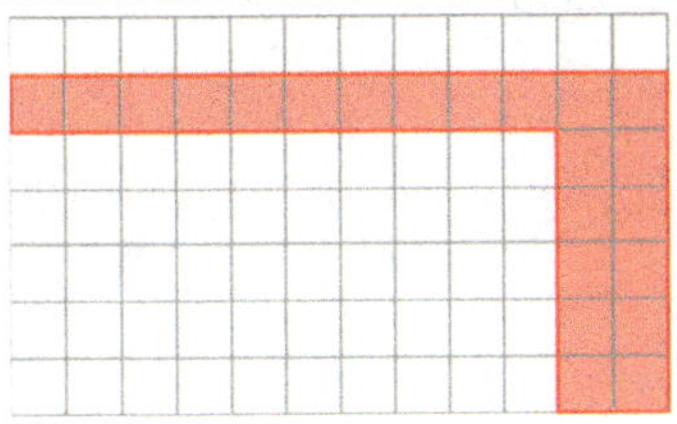

$10 \times 1 + 2 \times 6 = 22$ square units

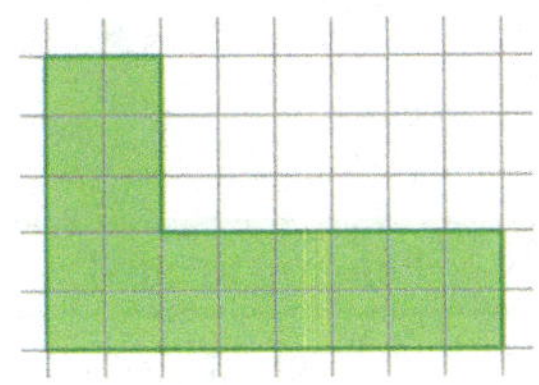

$2 \times 5 + 6 \times 2 = 22$ square units

4. The area is $48 \text{ ft} \times 48 \text{ ft} - 30 \text{ ft} \times 20 \text{ ft} = 2304 \text{ ft}^2 - 600 \text{ ft}^2 = 1704 \text{ ft}^2$.

5. a. $3 \text{ m} \times 3\text{m} + 7 \text{ m} \times 5 \text{ m} + 3 \text{ m} \times 3 \text{ m} = 9 \text{ m}^2 + 35 \text{ m}^2 + 9 \text{ m}^2 = 53 \text{ m}^2$.
   b. Divide the shape into two rectangles. That can be done in two ways. One way results in the calculation
   $16 \text{ ft} \times 32 \text{ ft} + 24 \text{ ft} \times 12 \text{ ft} = 512 \text{ ft}^2 + 288 \text{ ft}^2 = 800 \text{ ft}^2$.

Review: Area and Perimeter, pp. 13-16

1. a. perimeter   b. volume   c. area

2. a. perimeter = 32 in   area = 48 in^2
   b. perimeter = 200 cm   area = 2,500 cm^2

3. a. perimeter = 16 cm   area = 7 cm^2
   b. perimeter = 32 cm   area = 28 cm^2. Notice that when the perimeter gets doubled, the area gets *quadrupled*.

4. You can also use a letter or other symbol for the unknown, instead of ?.

   a. 9 ft + ? + 9 ft + ? = 32 ft. Solution: ? = 7 ft.

   b. 15 cm + ? + 15 cm + ? = 44 cm or 15 cm + ? = 22 cm. Solution: ? = 7 cm.

   c. ? + ? + ? + ? = 24 m or ? + ? = 12 m or 4 × ? = 24 m. Solution: ? = 6 m.

   d. 5 cm. The number sentence would be $s \times s = 25\ cm^2$. Its perimeter is 20 cm.

5.

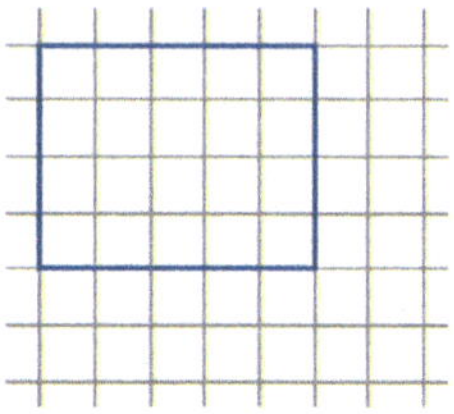

a. Sides 4 and 5 units; area 20 square units, perimeter 18 units.

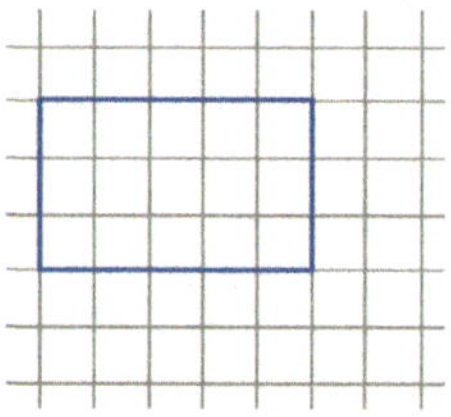

b. Sides 3 and 5 units; area 15 square units, perimeter 16 units.

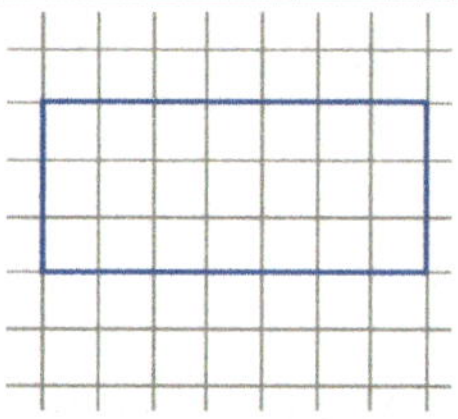

c. Sides 3 and 7 units; area 21 square units, perimeter 20 units.

6. 516 mm ÷ 6 = 86 mm

7. First divide the building into two rectangles. That can be done in two different ways. One way is shown on the right.
   a. The number sentence is then: 18 m × 33 m + 42 m × 18 m.

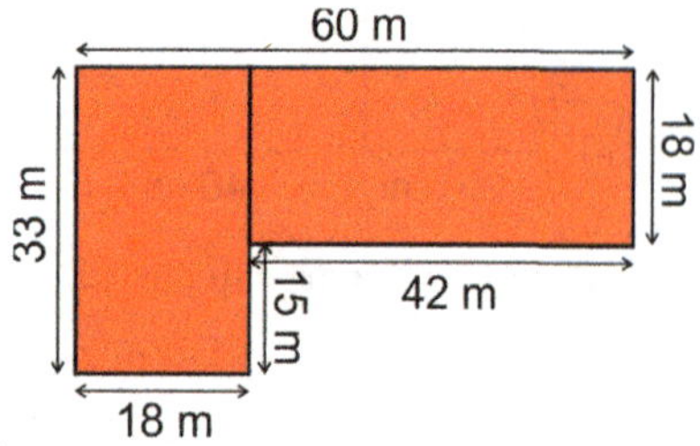

   b. The area is 18 m × 33 m + 42 m × 18 m
   = 594 m^2 + 756 m^2 = 1,350 m^2.

   c. The area of the whole plot of land is 80 m × 35 m = 2,800 m^2.
   We subtract the area of the school to get the area of the actual yard:
   2,800 m^2 − 1,350 m^2 = 1,450 m^2.

8. a. & b.

	Length	Width	Area	Fencing needed
Sheep yard 1	20 m	20 m	400 m^2	80 m
Sheep yard 2	10 m	40 m	400 m^2	100 m
Sheep yard 3	5 m	80 m	400 m^2	170 m

9. For example, 5 m by 10 m pen will work. Or, 6 m by 9 m. Or, 7 m by 8 m.

Puzzle corner. The width of the inside rectangle is 19 cm − 2 cm − 2 cm = 15 cm. The height of the inner rectangle is 14 cm − 2 cm − 2 cm = 10 cm. The area is 15 cm × 10 cm = 150 cm^2.

Lines, Rays, and Angles, pp. 17-19

1. a. Ray $\overrightarrow{AB}$ b. line $\overleftrightarrow{VW}$ c. line segment $\overline{MN}$

2.

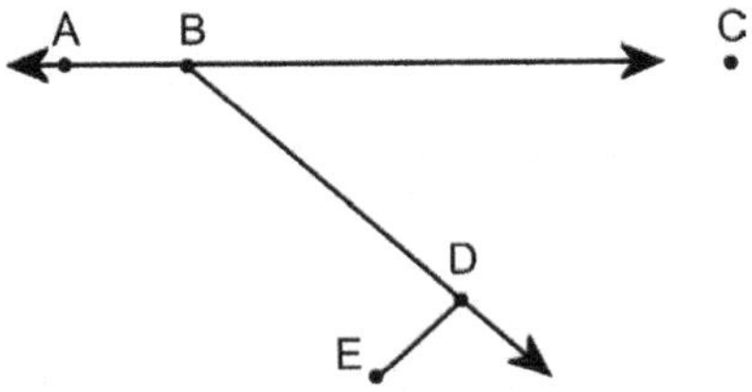

3. angle MAS or angle SAM

4. a. angle EDF or angle FDE
   b. angle ACE or angle ECA

5. Answers will vary. Check students' answers. For an example, see the image on the right.

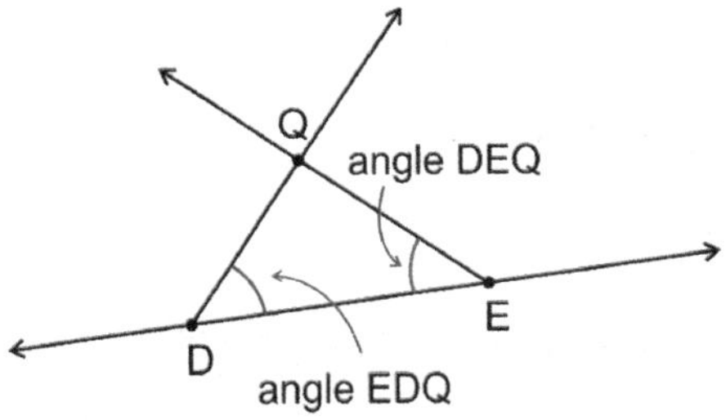

6. a. The second angle is bigger. b. the second angle c. the first angle
   d. the second angle e. the first angle f. the second angle

Measuring Angles, pp. 20-26

1. a. 35° b. 72° c. 18° d. 50°

2. a. 75° b. 100°
   c. 144° d. 135°
   e. 173° f. 93°

3. It is 180° −146° = <u>34°</u>.

4. a. 70° b. 45° c. 148° d. 125° e. 76° f. 107° g. 14°

5. a. acute b. right c. straight d. right e. acute f. obtuse g. acute h. acute i. obtuse

6. Check students' work. For example:

a.

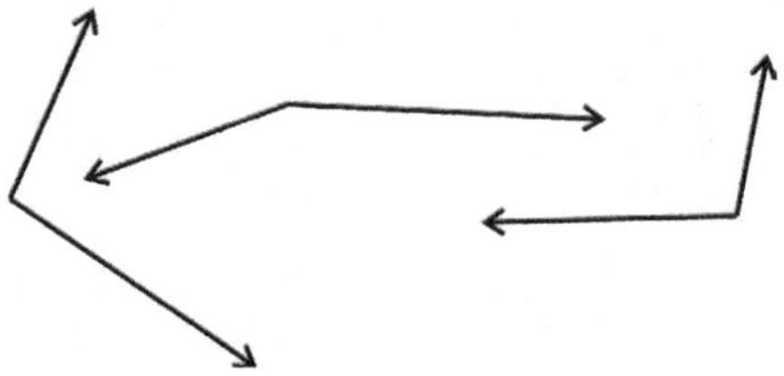

b.

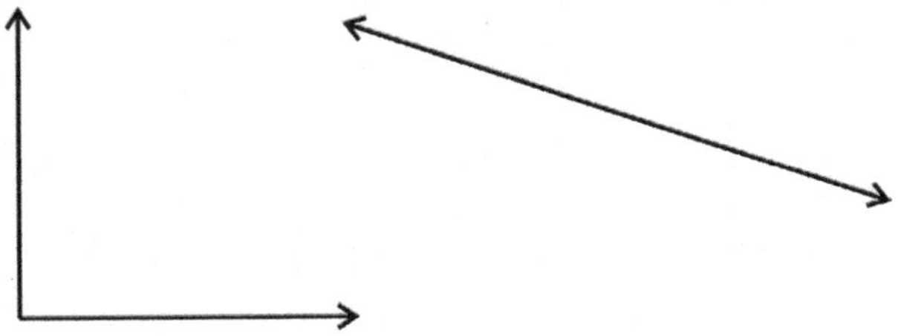

c.

7. The three angles measure 45° (acute), 102° (obtuse), 33° (acute).

Drawing Angles, pp. 27-28

1. Check the student's answers.

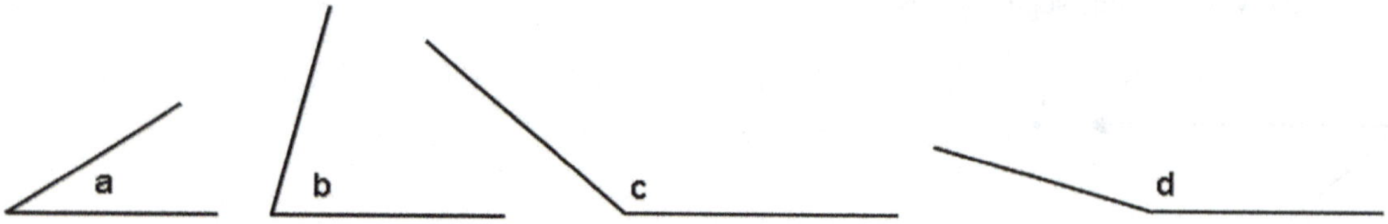

2. This is the general shape of the triangle with 85 and 25-degree angles:

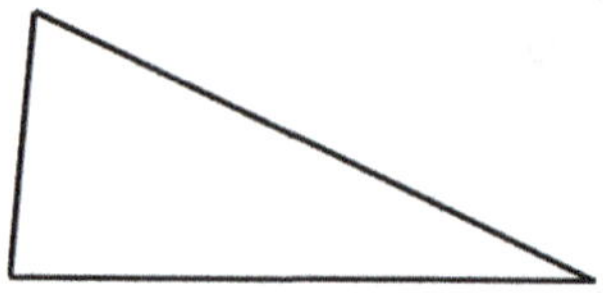

3. This is the general shape of the triangle with 30 and 60-degree angles. The student's triangle may also be a mirror image of this. Note: the third angle will be a right angle.

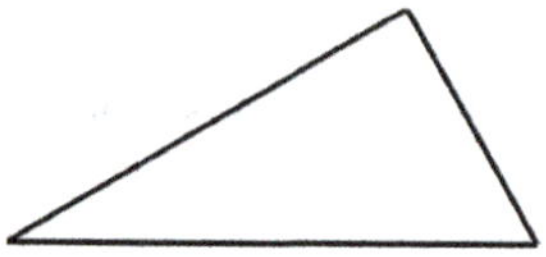

Estimating Angles, pp. 29-32

1. a. 60° b. 150°

2. Answers will vary. The following are the exact measurements.
   a. 30° b. 45° c. 100° d. 60° e. 160° f. 80°

3. Answers will vary. The following are the exact measurements.
   a. 40° b. 10° c. 15° d. 120° e. 270° f. 210°

4. Answers will vary. The following are the exact measurements: 60°, 30°, 90°.

5. Answers will vary. Check the student's answers.

6.

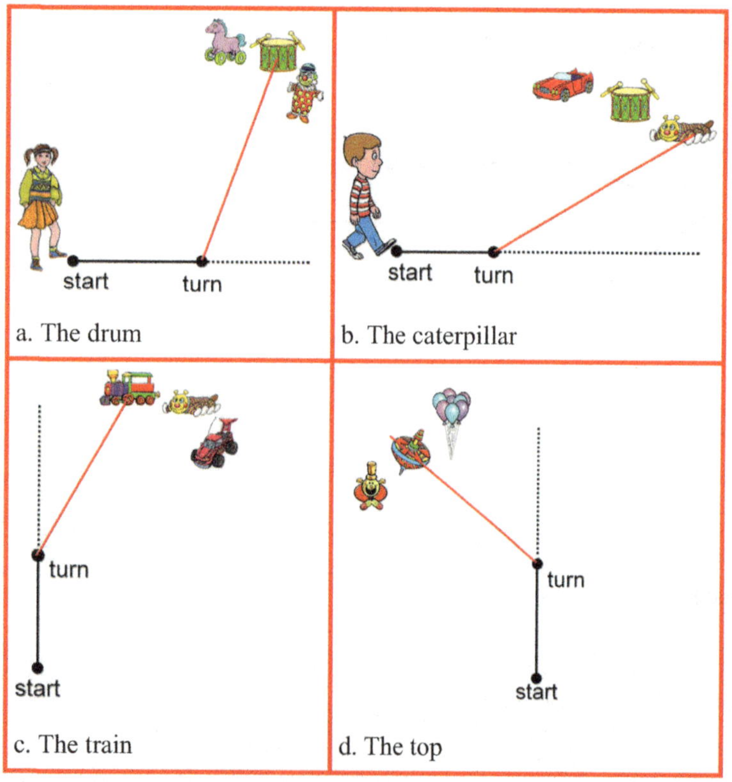

a. The drum
b. The caterpillar
c. The train
d. The top

Estimating Angles, continued, p. 33

7.

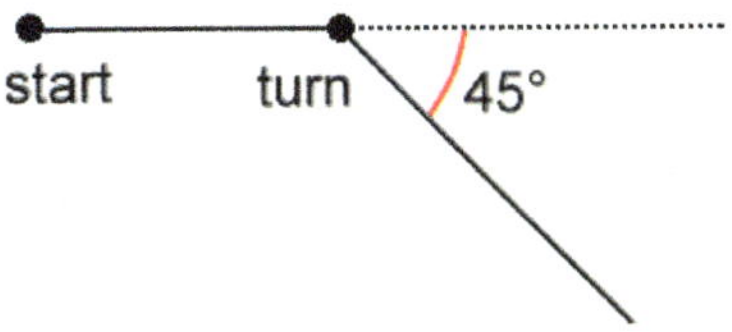

8.

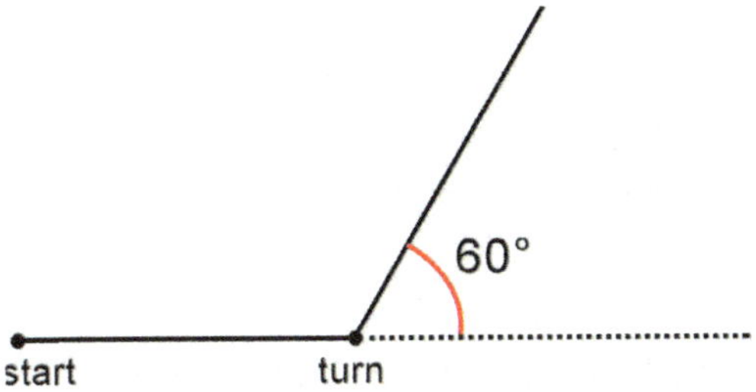

Puzzle corner: A SUPER TURNER

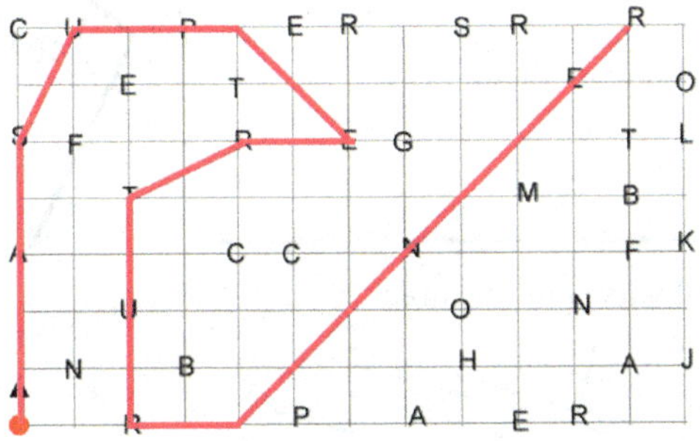

Angle Problems, pp. 34-36

1. Yes. You can subtract the angle measure of ∠DBC from 90° to get the measure of ∠ABD.

2. a. 74° b. 58° c. 65° d. 63°

3. a. 35° b. 45°

4. a. $38° + x = 72°$; $x = 34°$
   b. $37° + x = 90°$; $x = 53°$
   c. $47° + 23° + x = 122°$; $x = 52°$
   d. $34° + x = 105°$; $x = 71°$

Angle BAD is a straight angle, so it measures 180 °.

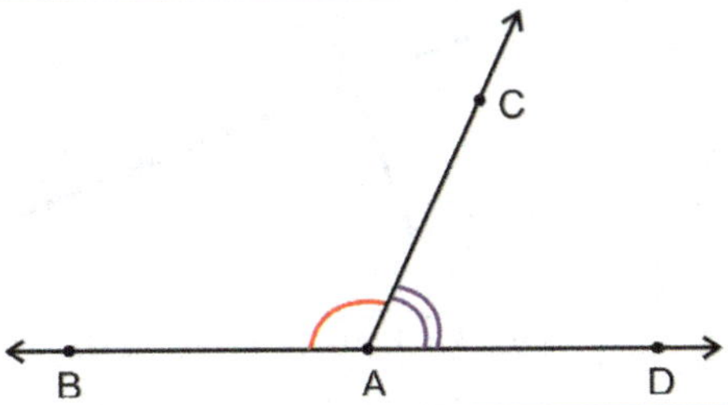

5. a. 68° b. 88°

6. a. $29° + x = 180°$; $x = 151°$
   b. $x + 54° = 180°$; $x = 126°$

Angle Problems, continued, pp. 37-38

There are THREE angles in this picture that share the same vertex. Notice that they three make a complete circle!

Measure them. Find the sum of the angle measures.

Angle A: 155 °

Angle B: 80 °

Angle C: 125 °

Sum: 360 °

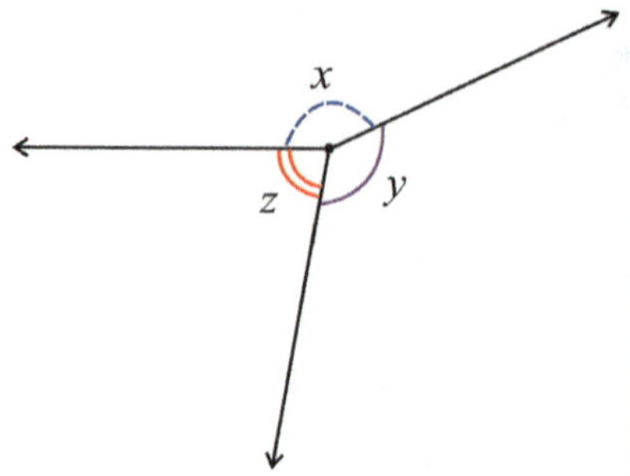

7. a. $28° + 116° + 54° + x = 360°$; $x = 162°$
   b. $57° + 113° + x = 360°$; $x = 190°$.

8. The sum of all three angle measures is 300°. (The first angle measures 180°, the second 90°, and the last one 30°.)

9. Measure first the smaller angle that you can measure with the protractor, marked with a dashed line. It is 102°. Then subtract that from 360°: $360° - 102° = 258°$. That is the measurement of the angle in question.

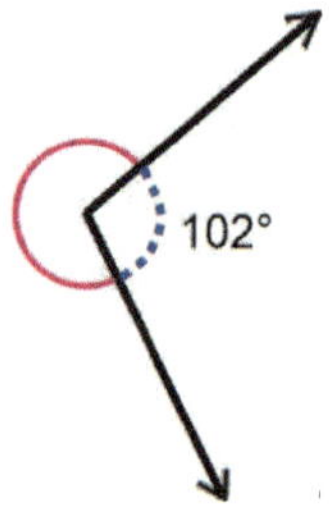

Puzzle corner. a. It is $360° \div 8 = 45°$. b. 120°. (First divide 360° by six, then double that.)

Parallel and Perpendicular Lines, pp. 39-43

1. They are not parallel. They intersect.

2. a. Line segments AB and BC are perpendicular.
   Line segments AD and BC are parallel.

   b. Line segments EF and GH are parallel.
   Line segments EH and FG are parallel.

3.

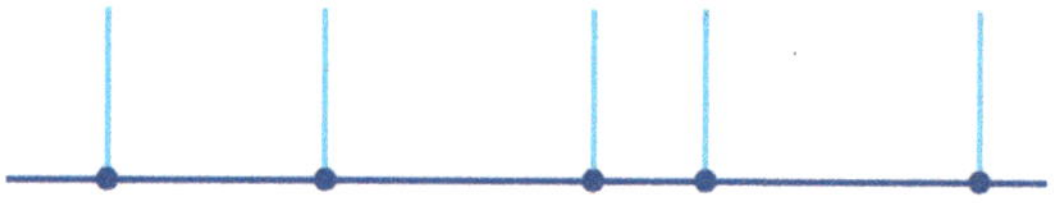

4.

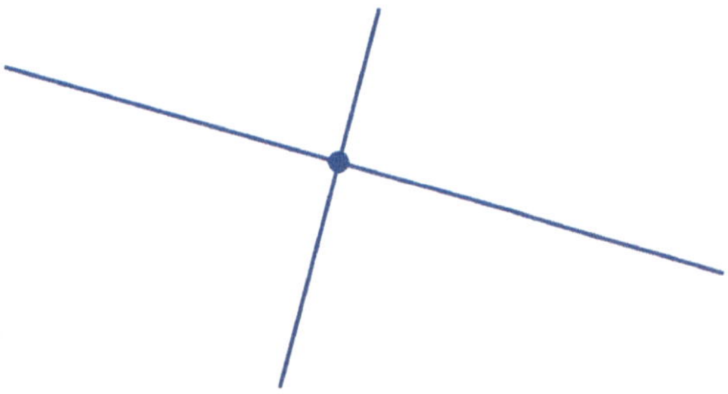

5.

a.

b.

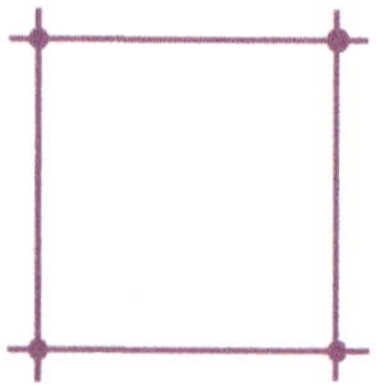

6. Answers will vary. Check students' answers.

7. Answers will vary. For example:

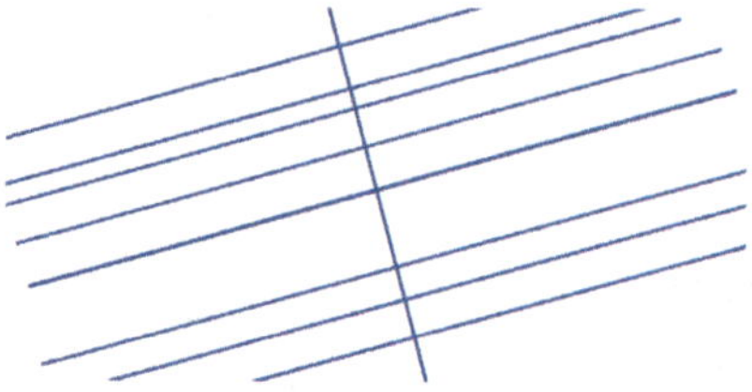

8. Answers will vary. Check the student's answer. For example:

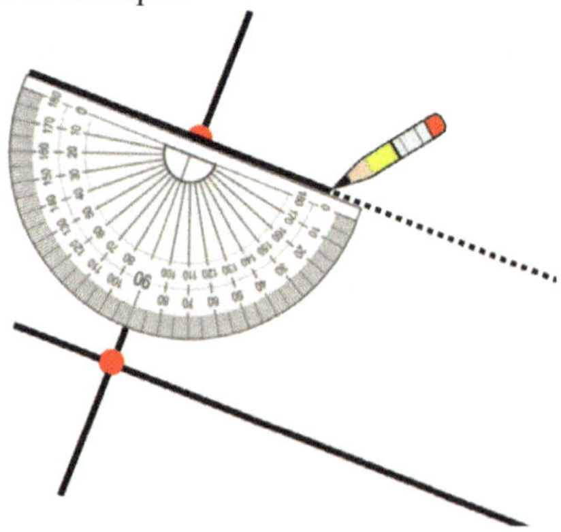

9. $\overline{AF} \perp \overline{FB}$ $\quad$ $\overline{AF} \parallel \overline{BE}$

   $\overline{FB} \perp \overline{BE}$ $\quad$ $\overline{AB} \parallel \overline{FE}$

10. a. $\overline{AB} \perp \overline{AD}$, $\overline{AD} \perp \overline{CD}$, $\overline{AB} \parallel \overline{CD}$

    b. $\overline{AB} \perp s$, $r \perp t$, $t \perp u$, $r \parallel u$

    c. $s \perp t$, $s \perp \overline{BC}$, $s \perp \overline{FE}$,
    $\overline{AF} \parallel \overline{CD}$, $\overline{AB} \parallel \overline{DE}$, $\overline{BC} \parallel \overline{EF}$, $\overline{EF} \parallel t$, $\overline{BC} \parallel t$.

Parallelograms, pp. 44-46

1. a. Answers will vary. For example:

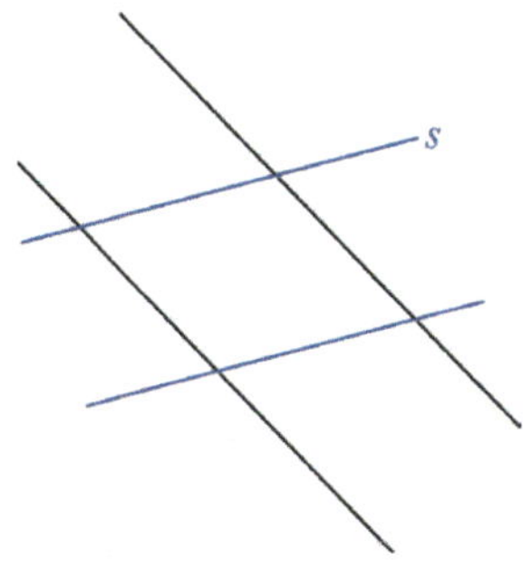

b. Answers will vary, but the opposite sides should measure the same.

2. Answers will vary. Check the student's answers. The opposite sides should measure the same (be congruent), as also should the opposite corners.

3. Answers will vary. Check the student's answers. For example:

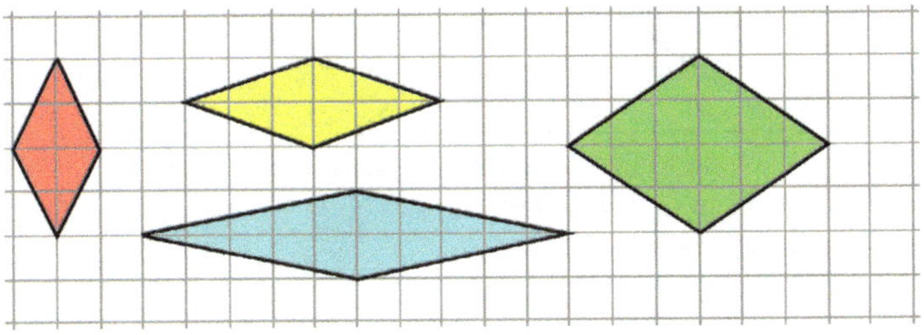

4. Answers will vary. Check the student's answers. For example:

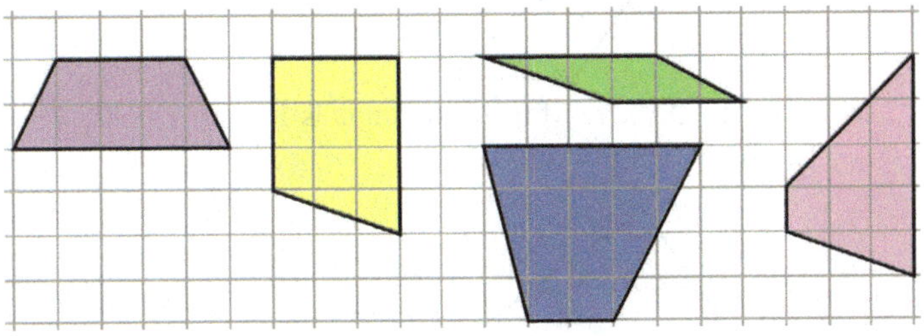

5. a. trapezoid   b. parallelogram   c. trapezoid   d. rhombus

6. a. The image is not to scale:

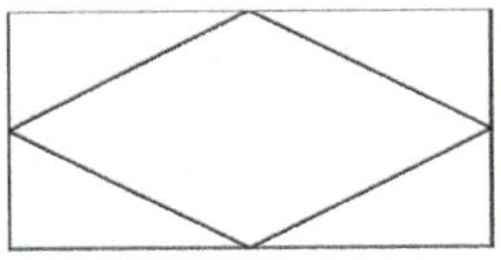

b. a rhombus. (It is also a parallelogram.)

7. A: trapezoid   B: parallelogram   C: trapezoid
D: rectangle   E: rhombus   F: (no special name or a *scalene quadrilateral*)

Puzzle corner:
Yes. A rectangle is a parallelogram, because its opposite sides are parallel.
A square is a parallelogram, too.
A square is a rhombus, yes.
The rectangle in the question is not a rhombus.

Triangles, pp. 47-50

1. a. and b. Use a protractor or a triangular ruler to draw the right angle.
   c. The other two angles in a right triangle are acute.

> **A right triangle has one right angle.**
> **The other two angles are acute.**

2. c. In an obtuse triangle the other two angles are acute.

> **An obtuse triangle has one obtuse angle.**
> **The other two angles are acute.**

3. a. b. Answers will vary. Check the student's answers. All the angles should be acute (less than 90°).

4.

> **Right triangles** have exactly one right angle, and the other two angles are acute.
>
> **Obtuse triangles** have exactly one obtuse angle, and the other two angles are acute.
>
> **Acute triangles** have three acute angles.

5. a. obtuse
   b. right
   c. acute
   d. obtuse
   e. acute
   f. right
   g. Triangle ABD: right.
   Triangle ACD: obtuse.
   Triangle BCD: right.

6. a. acute
   b. right
   c. obtuse
   d. obtuse
   e. The black triangle is obtuse.
   The red triangle is acute.

7. a. 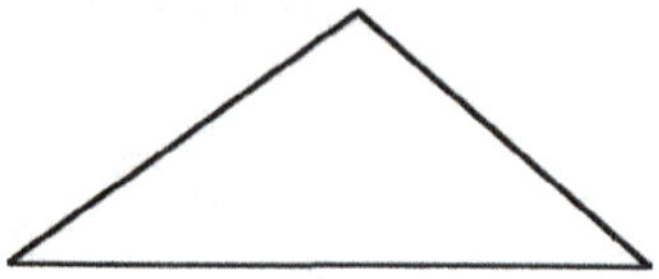
   b. The measure of the third angle is 105°.
   c. It is an obtuse triangle.

8. a. 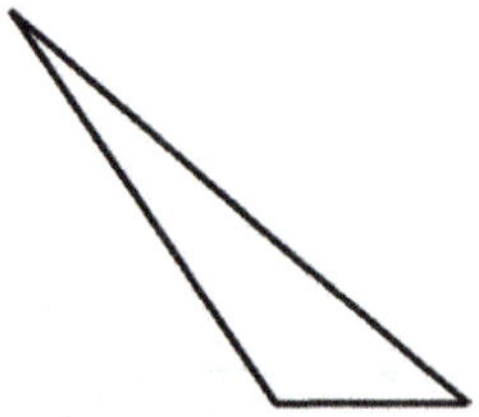
   b. The measure of the third angle is 15°.
   c. It is an obtuse triangle.

9. a. 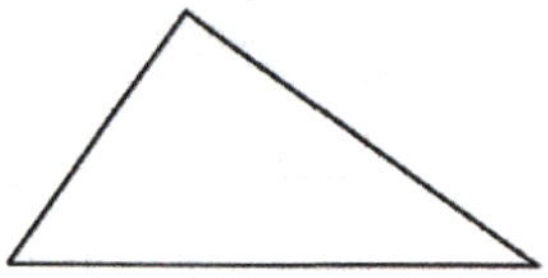
   b. The measure of the third angle is 90°.
   c. It is a right triangle.

Line Symmetry, pp. 51-52

1. a. no b. yes c. yes d. no e. no f. no g. no h. yes i. yes

2. a. 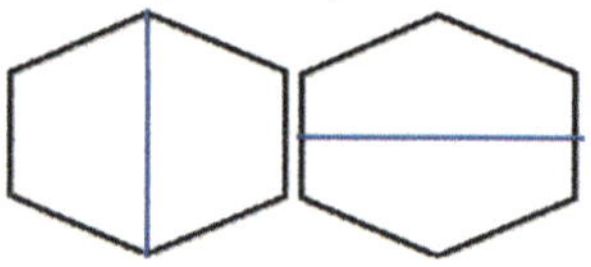 b. 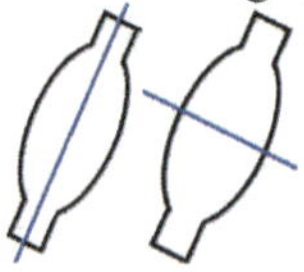 c. 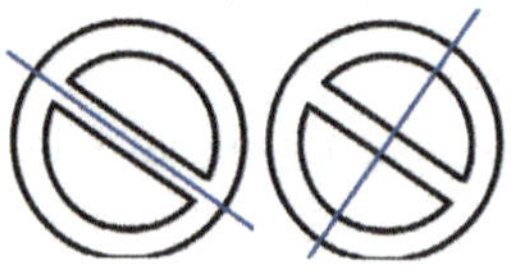 d. 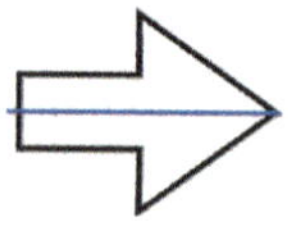

e. 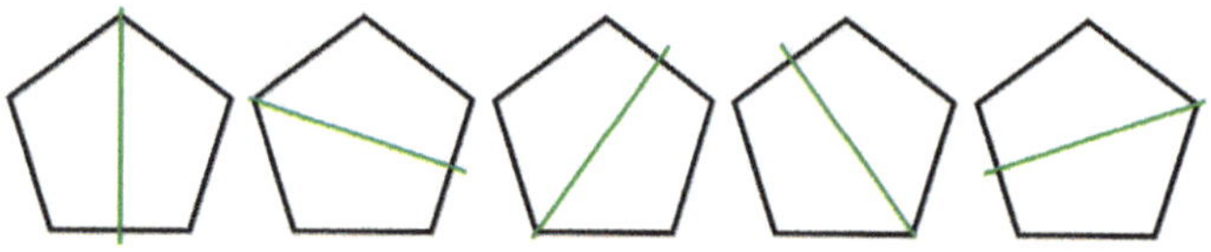

f. 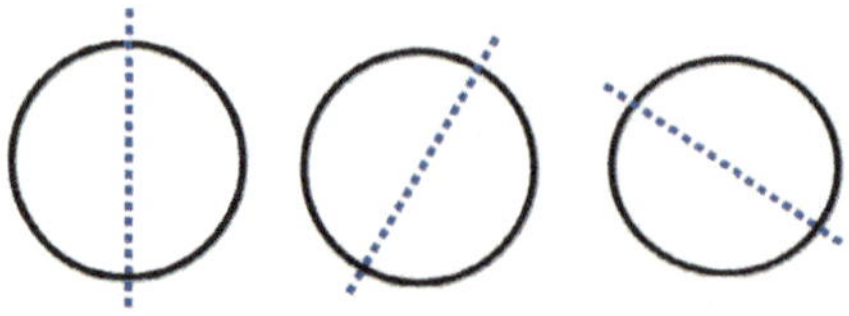

...and many more. Any diameter of a circle (a line through its center point) is its symmetry line.

3. You can draw a vertical symmetry line in the letters A, H, I, M, O, T, U, V, W, X, and Y.
   You can draw a horizontal symmetry line in the letters B, C, D, E, H, I, K, O, and X.

Line Symmetry, continued, p. 53

4. a. 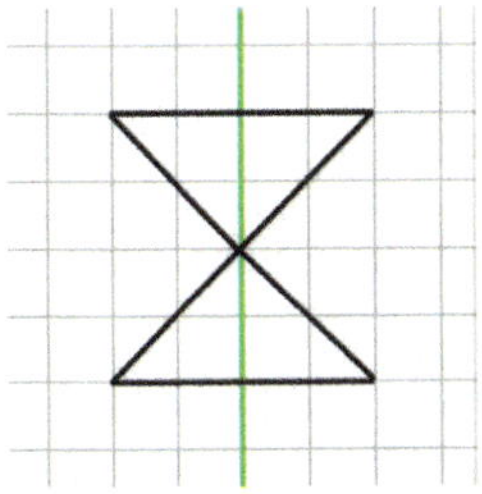 b. 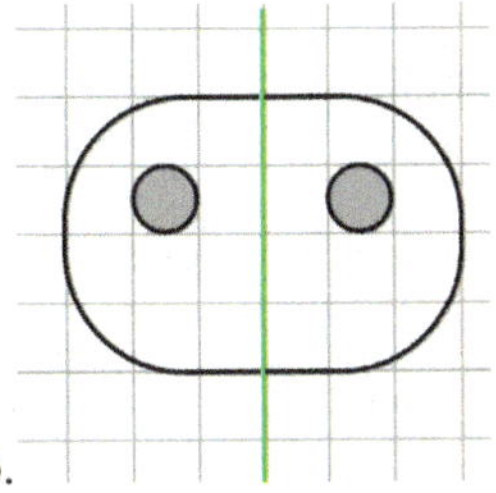 c. 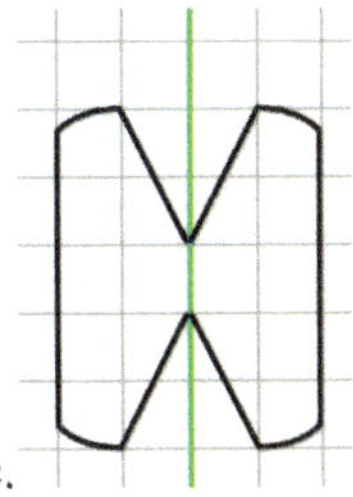 d. 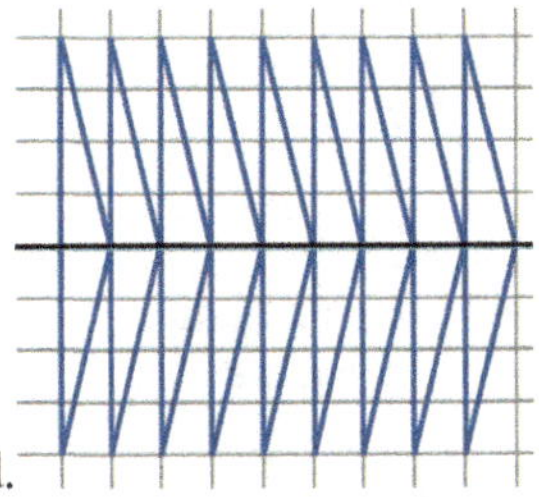

Review: Angles, pp. 54-57

1. a. 90° b. 27°

2. a. b. Answers will vary. Check the student's answers. The acute angle should measure less than 90°, and the obtuse angle more than 90°.

3. Answers will vary. Check the student's triangle and the classification of the angles.

4. The 76°-angle may also be on the left side, instead of on the right, like in the image below.

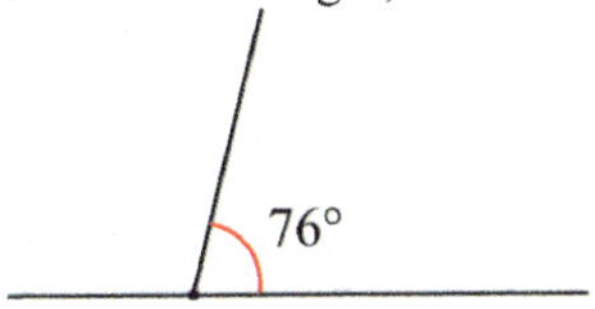

5. a. 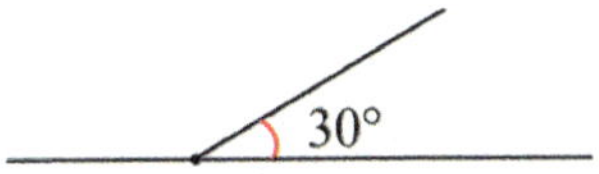

b. 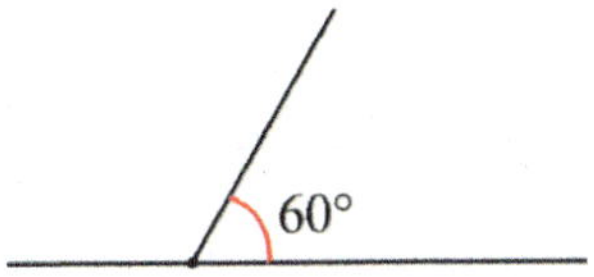

c. 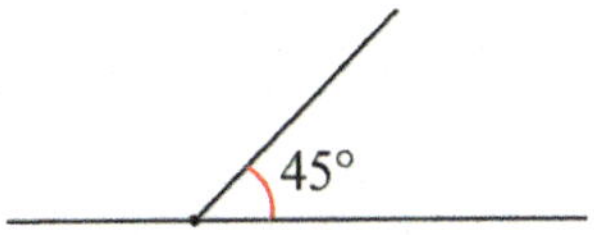

6. a. 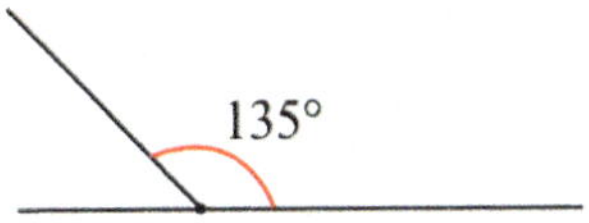

b. 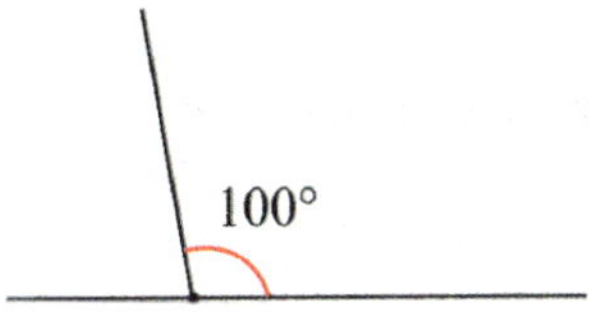

c. 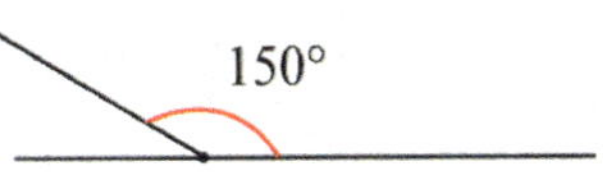

7. Estimates will vary. The exact measures are:
   a. 15° b. 94° c. 55° d. 121°

8.

a. 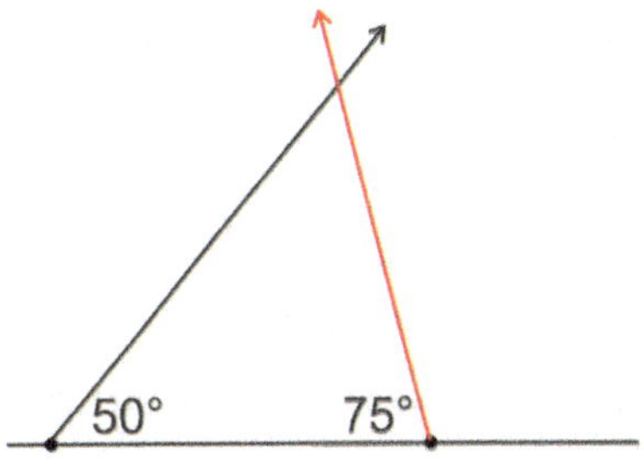

b. The third angle measures 55°. What the student measures may be slightly different from that, depending on how accurate the drawing is.

c. All three angles are acute.

9.

a. 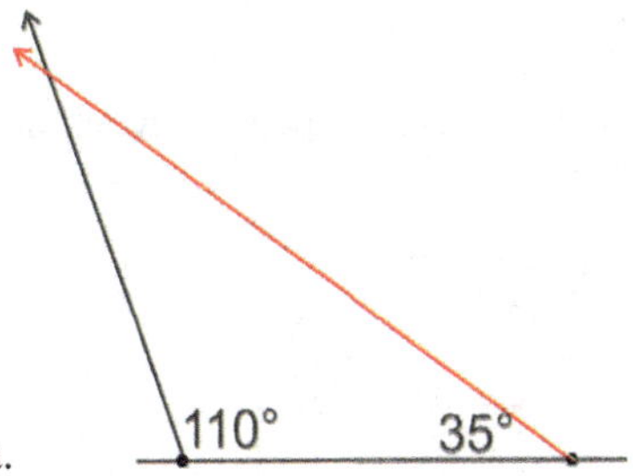

b. The third angle measures 35°. What the student measures may be slightly different from that, depending on how accurate the drawing is.

c. The 110° angle is obtuse; the other two are acute.

Polygons, pp. 58-60

1. Polygons: b, c, d, j, k.
   Not polygons: a, e, f, g, h, i.

2. Square.

3. a. regular pentagon b. pentagon
   c. regular heptagon d. square
   e. octagon f. hexagon
   g. quadrilateral h. hexagon

4. a. 135° b. 108°

5. Things that students notice will vary. Several possibilities are listed below.

   a. Angle A = 58°; Angle B = 122°;
   Angle C = 58°; Angle D = 122°.

   - I notice that angles A and C are congruent, and also angles B and D.
   - I notice that angles A and B (or C and D) add up to 180°.
   - I notice that the sum of A, B, C, and D is 360°.
   - I notice that A and C are acute, and B and D are obtuse.

5. b. Angle A = 69°; Angle B = 111°;
   Angle C = 111°; Angle D = 69°.

   - I notice that angles A and D are congruent, and also angles B and C.
   - I notice that angles A and B (or C and D) add up to 180°.
   - I notice that the sum of A, B, C, and D is 360°.
   - I notice that A and D are acute, and B and C are obtuse.

Puzzle corner.
A = 60°, B = 120°, C = 60°.

Classifying Quadrilaterals 1, pp. 61-63

1.

No parallel sides	One pair of parallel sides	Two pairs of parallel sides
c. d. h.	b. e. i.	a. f. g. j.

2. Answers vary. The example answer below has several possible figures for b, c, and d.

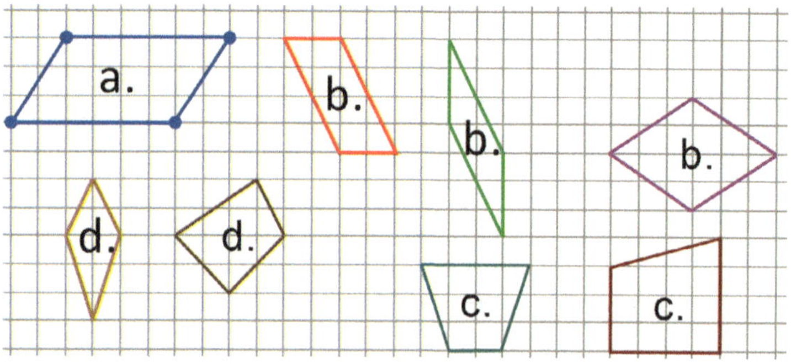

3. a. Yes. A rectangle has two pairs of parallel sides, so it is a parallelogram.

   b. Yes. The definition of a trapezoid says that it has *at least* one pair of parallel sides. A parallelogram has *two pairs* of parallel sides, so it fulfills the definition of a trapezoid.

   c. No. For example, this trapezoid  is not a parallelogram; it doesn't have two pairs of parallel sides.

   d. Yes. All rectangles are parallelograms, and all parallelograms are trapezoids, so all rectangles are trapezoids. However, the opposite is not true. Not all trapezoids are rectangles.

4. Because all parallelograms are trapezoids; a parallelogram fulfills the definition of a trapezoid (see the answer to question 3b above).

5. They would go in the set of parallelograms.

Classifying Quadrilaterals 1, cont., p. 63

6.

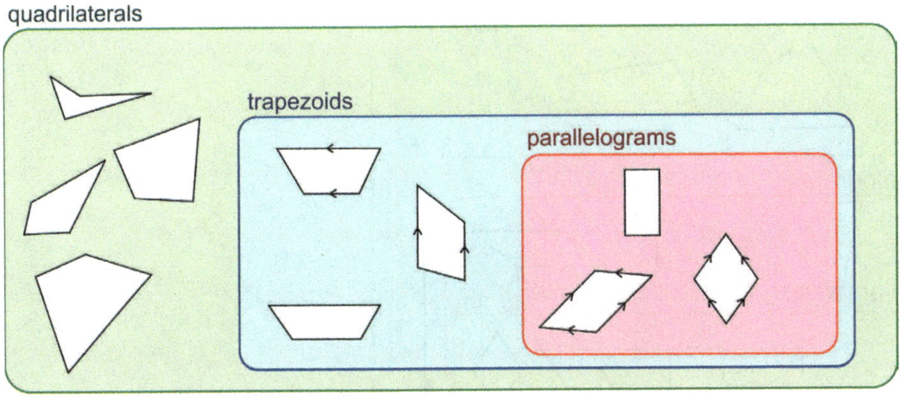

Classifying Quadrilaterals 2, pp. 64-65

1. Yes, it does. A square has four congruent sides (and also two pairs of parallel sides).

2. a. The <u>rectangle</u> is like a "child" to the parallelogram: it *is* a parallelogram. Additionally, its angles are right angles.

   b. The square is like a "child" to the <u>rectangle</u> : it, too, has four right angles. Additionally, <u>it has four congruent sides.</u>

3. It would go under the parallelogram, but above the square:

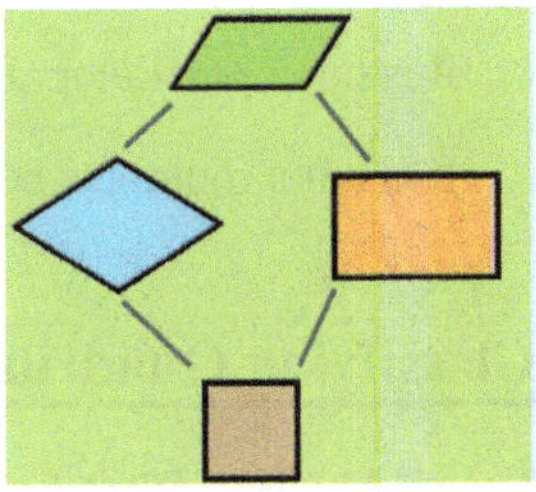

   This is because a rhombus *is* a parallelogram (it has two pairs of parallel sides). And since it has four congruent sides, which a square also has, it is like a "parent" to the square. (The square has an additional characteristic: its angles are right angles.)

4. The squares go in the overlapping area of the rhombi and rectangles.

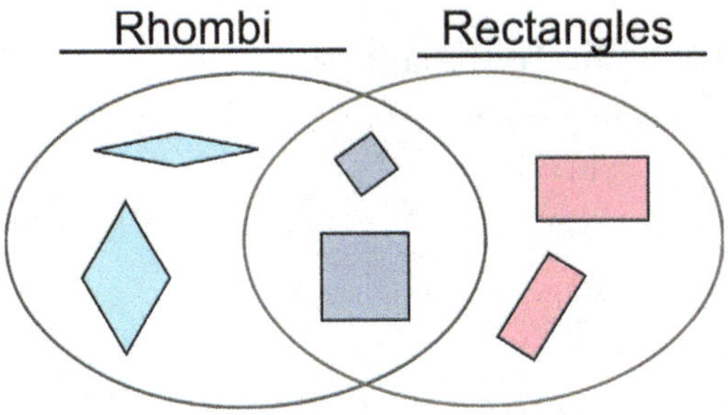

5. a. Yes. A rhombus also has two pairs of congruent sides, like a kite. The two pairs just happen to be congruent.
   b. No. A rectangle's adjacent sides are of different lengths—unless the rectangle is also a square.
   c. Yes. A square has two pairs of congruent sides, like a kite. Same as the rhombus, the two pairs are congruent.
   d. No—unless the trapezoid in question happens to be a square. A trapezoid doesn't have to have any congruent sides.

6. a. parallelogram   b. kite
   c. trapezoid   d. rhombus
   e. trapezoid   f. parallelogram
   g. kite   h. trapezoid
   i. rhombus   j. parallelogram

Classifying Quadrilaterals 2, cont., p. 66

7. 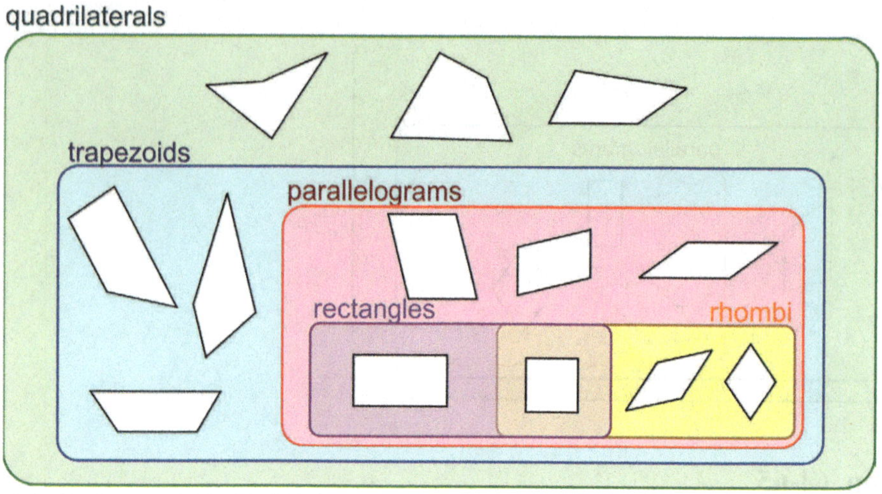

8. She can draw several different kinds: not only a square, but also rhombi of various shapes that are not squares.

9. It is a parallelogram. A parallelogram has two pairs of congruent sides by definition. And the opposite sides of a parallelogram are congruent.

10. It is not a kite. A kite has two pairs of congruent sides, such as 3 cm, 3 cm, 8 cm, and 8 cm. The lengths given in the exercise only have one pair of congruent sides (3 cm and 3 cm).

    It is not a parallelogram, because in a parallelogram, the opposite sides are congruent (the same length).

    It is not a rhombus, because in a rhombus, all sides are congruent.

Classifying Quadrilaterals 3, pp. 67-68

1. You can draw several. You can draw a square with 2-in sides, or you can draw various rectangles where the one side is 2 inches and the other side is longer or shorter.

2. It is not a kite nor a rhombus. The adjacent (neighboring) sides of this parallelogram are not congruent, so it cannot be a kite or a rhombus. However, it *is* a trapezoid. Every parallelogram fulfills the definition of a trapezoid, because a parallelogram has *at least* one pair of parallel sides.

3. Yes, you can, by varying how tall the trapezoid is or by varying the angles. For example (the figures are not to scale):

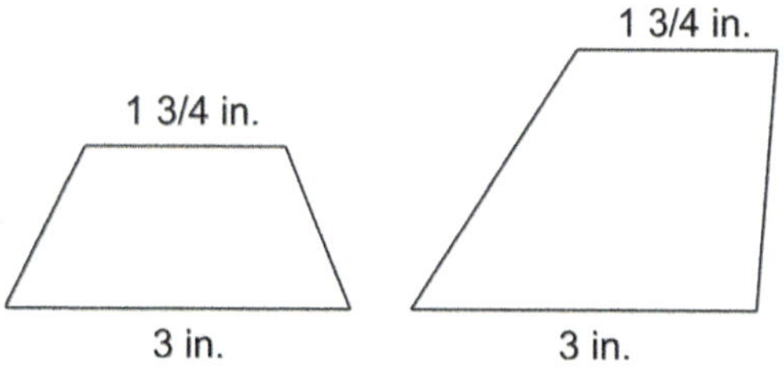

4. 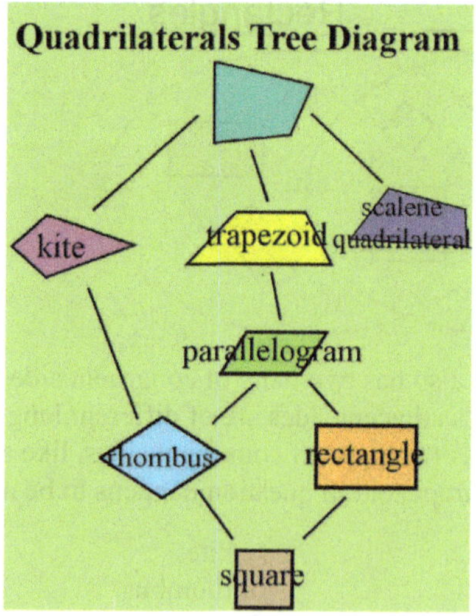

5. a. Yes. b. Yes. c. No. d. Yes. e. No.

6. Quadrilaterals puzzle:
   YOU GOT THE SHAPES SORTED!

Classifying Triangles 1, pp. 69-71

1. Check the student's sorting. They may sort them according the types of angles (acute, right, obtuse), or according to how many congruent sides the triangle has (none, two, three), or according to some other characteristic.

   There are no right or wrong answers for this exercise; it is to get the student(s) to thinking.

2. Equilateral: f, g.
   Isosceles: a, c, d (and f and g)
   Scalene: b, e, h.

3. 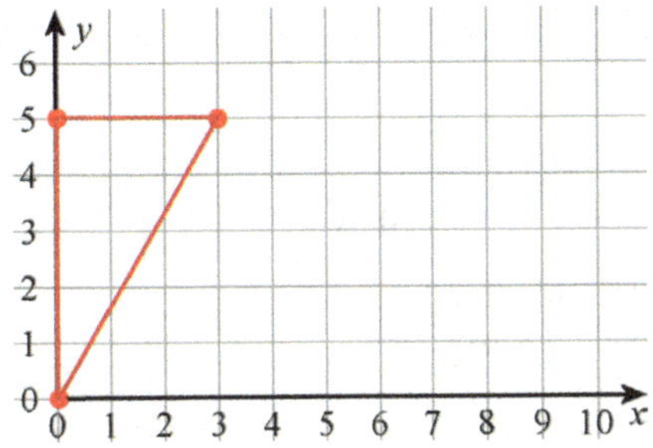

   It is a scalene triangle.

4. a. obtuse b. acute c. right d. acute e. acute
   f. right g. obtuse h. right

5. Because a triangle can only have one right angle.

   IF a triangle had two right angles, the two sides (marked with arrows in the image) would never meet... it would not be a triangle.

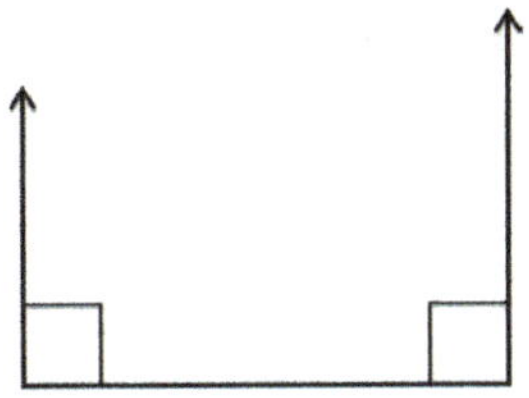

6. 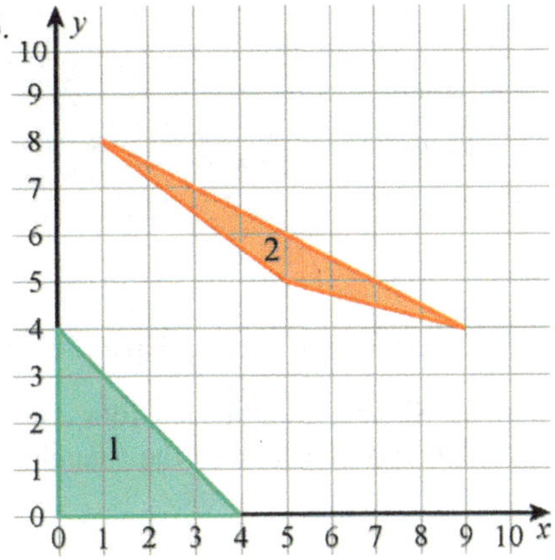

   Triangle 1: right and isosceles
   Triangle 2: obtuse and scalene

Classifying Triangles 2, pp. 72-73

1.

Triangle	Classification by the angles	Classification by the sides
a.	acute	equilateral
b.	right	scalene
c.	obtuse	isosceles
d.	acute	scalene

2. Yes. Yes. Yes. The sketches will vary. Check the student's answer. For example:

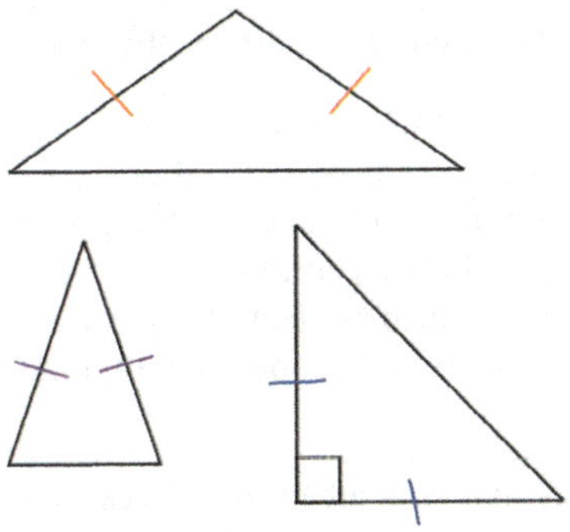

3. Yes. Yes. Yes. The sketches will vary. Check the student's answer. For example:

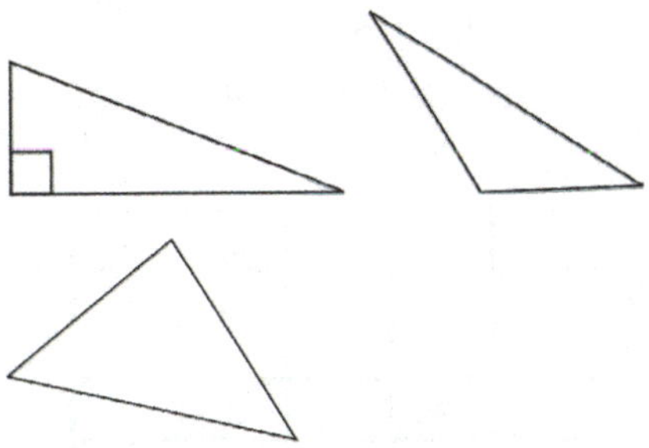

4. Answers will vary. Check the student's answer. For example:

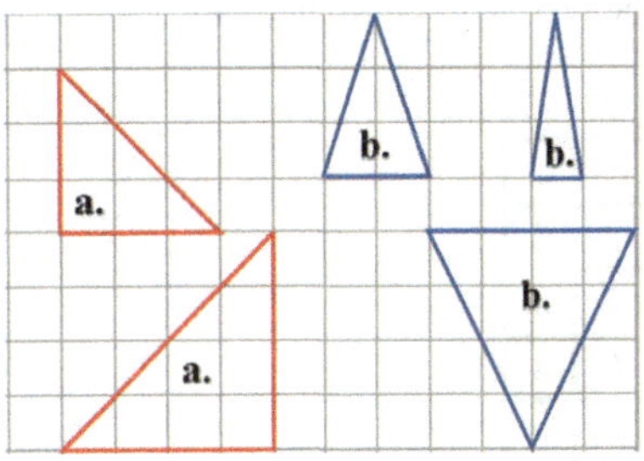

Classifying Triangles 2, cont., p. 73

Page 195

5.

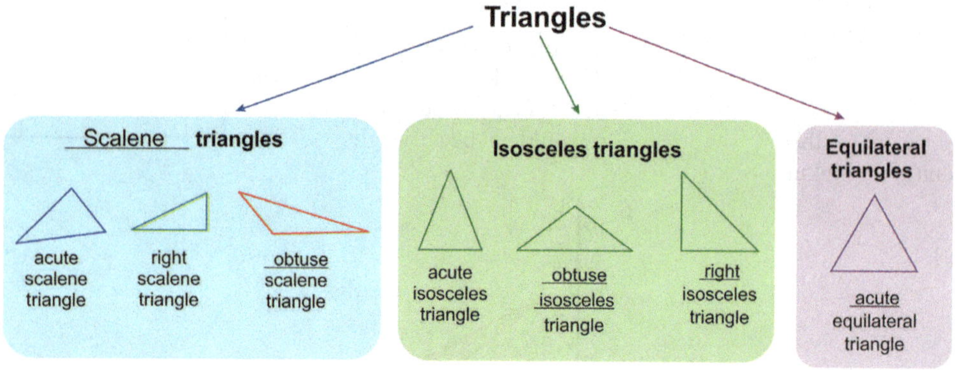

6. The angles measure 60°.

7. Answers will vary; check the student's sketch. A parallelogram can work well for this. For example:

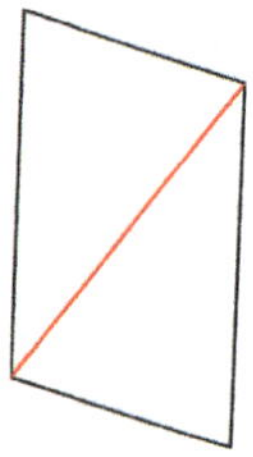

Puzzle corner. It is a rhombus. Since a rhombus has four congruent sides, when you divide it into two triangles with a diagonal, two of the sides of the triangle will be congruent—hence it's isosceles.

Volume, pp. 74-78

1. a. V = 6 cubic units b. V = 16 cubic units
   c. V = 12 cubic units d. V = 32 cubic units
   e. V = 36 cubic units f. V = 28 cubic units
   g. V = 12 cubic units h. V = 20 cubic units
   i. V = 16 cubic units

2. a. V = 3 in^3 or 3 cu. in. b. V = 30 ft^3 or 30 cu. ft.
   c. V = 12 cm^3 d. V = 6 m^3
   e. V = 13 cm^3 f. 19 in^3 or cu. in.

3.

	a.	b.	c.	d.
Cubes in the bottom layer	*8*	12	21	6
Height	*4*	4	3	6
Volume	*32*	48	63	36

4. There are 10 × 3 × 3 = 90 cubes that fit into the box. So, the volume of the box is 90 cubic inches.

5.

	a.	b.	c.	d.
Width:	3	3	3	4
Depth:	3	2	2	4
Height:	2	3	6	4
Volume:	18	18	36	64

6. a. V = 3 × 3 × 1 = 9 cubic units
   b. V = 3 × 2 × 2 = 12 cubic units
   c. V = 3 × 2 × 5 = 30 cubic units
   d. V = 4 × 4 × 3 = 48 cubic units

7. She needs 4 × 4 × 4 = 64 more cubes to fill the box.

8. Check students' answers. The orientation of the prisms may vary. For example:

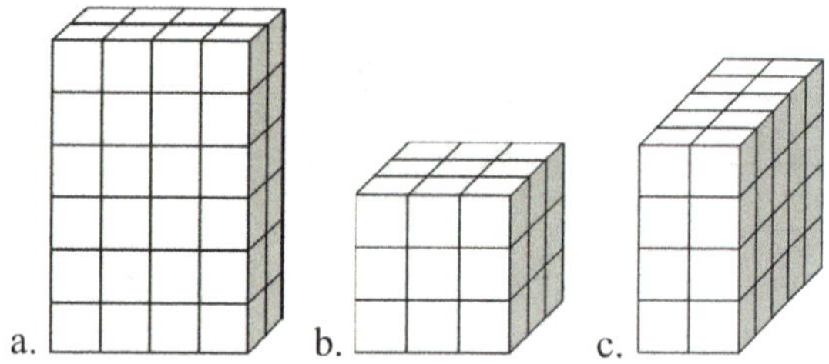

9. This question can be answered in two different ways.

   Some students might say that Chris is wrong and Mia is correct. This is a correct answer based on how the lesson presented the volume of a rectangular prism as width × depth × height. In this case, you might point out that Chris's prism could also be seen as correct, if the order of the dimensions (width, depth, height) has not been specified.

   Some students may say that both are correct, because Chris's prism is just in a different orientation; it still represents a prism with dimensions of 5, 3, and 2 units.

10. c. 2 × 2 × 2 + 2 × 2 × 5

Volume of Rectangular Prisms, pp. 79-81

1. The order of factors in the multiplications may vary.

   a. $V = \underline{5\text{ in}} \times 7\text{ in} \times 2\text{ in} = 70\text{ in}^3$

   b. $A = 25\text{ ft} \times 20\text{ ft} = 500\text{ ft}^2$

   $V = 25\text{ ft} \times 20\text{ ft} \times 9\text{ ft} = 4{,}500\text{ ft}^3$

2. a. $20\text{ cm} \times 30\text{ cm} \times 60\text{ cm} = 36{,}000\text{ cm}^3$

   b. $16\text{ in}^2 \times 6\text{ in} = 96\text{ in}^3$

3. Its depth is 2 cm, because $5\text{ cm} \times 3\text{ cm} \times 2\text{ cm} = 30\text{ cm}^3$.

4. Check the student's answers. The measurements can be done in inches, in centimeters, or meters, and the volume will be correspondingly in cubic inches, cubic centimeters, or cubic meters.

5. Sketches will vary. Check the student's answer. Its volume is $6\text{ cm} \times 5\text{ cm} \times 4\text{ cm} = 120\text{ cm}^3$.

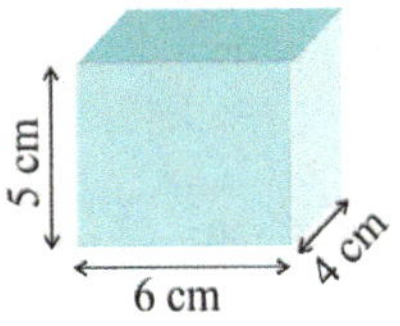

6. Sketches will vary. Check the student's answer. Its volume is $50\text{ cm} \times 80\text{ cm} \times 90\text{ cm}$ $= 360{,}000\text{ cm}^3$.

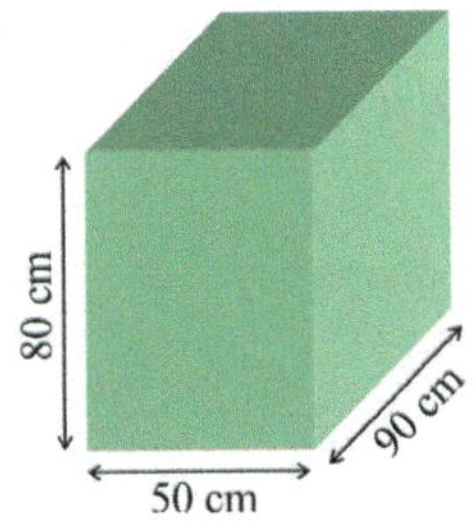

7. Its height is 6 inches, because $5\text{ in} \times 6\text{ in} \times 6\text{ in} = 180\text{ in}^3$. Or, you can use division and solve $180 \div (5 \times 6) = 6$.

8. The volume of the entire aquarium is $40\text{ cm} \times 27\text{ cm} \times 22\text{ cm} = 23{,}760\text{ cm}^3$. Since it is 1/4 full, the volume of water is $23{,}760\text{ cm}^3 \div 4 = \underline{5{,}940\text{ cm}^3}$.

9. a. Answers vary. Check the students' work. One possibility is: Width 4 in, height 4 in, and depth 4 in. Another is: Width 2 in, height 8 in, and depth 4 in.

   b. Answers vary. Check the student's work. One possibility is: 80 cm by 120 cm by 100 cm. Another is: 100 cm by 100 cm by 96 cm.

10. a. It is 4×18 cubic units = 72 cubic units.

    b. Answers vary. Check the students' work. For example, it could be 6 by 6 by 2 units, or 2 by 4 by 9 units, or 6 by 3 by 4 units, and so on.

11. a. $12\text{ ft} \times 18\text{ ft} \times 8\text{ ft} = 1{,}728$ cu. ft.

    b. The room loses $12\text{ ft} \times 18\text{ ft} \times 1\text{ ft} = 216$ cubic feet of volume.

Puzzle corner. A cubic yard is $3\text{ ft} \times 3\text{ ft} \times 3\text{ ft} = 27\text{ ft}^3$, so two cubic yards is 54 ft^3.

Volume Is Additive, pp. 82-84

1. $V_1 = 20\text{ cm} \times 20\text{ cm} \times 20\text{ cm} = 8{,}000\text{ cm}^3$

   $V_2 = 50\text{ cm} \times 20\text{ cm} \times 20\text{ cm} = 20{,}000\text{ cm}^3$

   $V_{total} = 20{,}000\text{ cm}^3 + 8{,}000\text{ cm}^3 = 28{,}000\text{ cm}^3$

   Notice that the depth of the top box is the same as the depth of the bottom box (20 cm). The total volume is found by adding and is $28{,}000\text{ cm}^3$.

2. a.

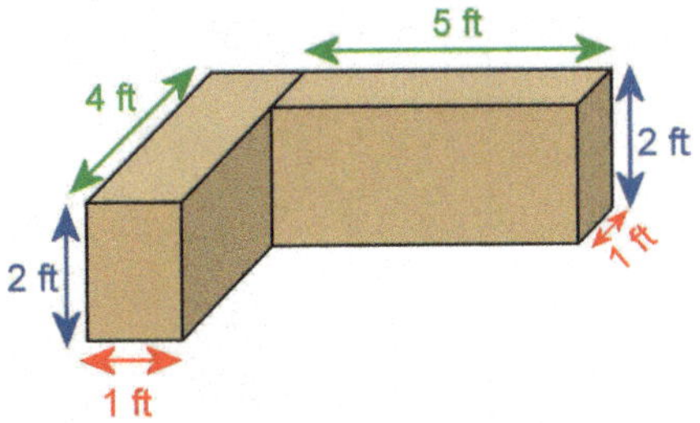

   b. The volume is $1\text{ ft} \times 4\text{ ft} \times 2\text{ ft} + 1\text{ ft} \times 5\text{ ft} \times 2\text{ ft}$ $= 8\text{ ft}^3 + 10\text{ ft}^3 = 18\text{ ft}^3$.

3. The total volume is $\underline{936\text{ m}^3}$. The volume of the "bottom" part is $12\text{ m} \times 8\text{ m} \times 6\text{ m} = 576\text{ m}^3$. The volume of the upper part is $5\text{ m} \times 8\text{ m} \times 9\text{ m} = 360\text{ m}^3$. The total volume is $576\text{ m}^3 + 360\text{ m}^3 = 936\text{ m}^3$.

4. The volume of the lowest part: $12\text{ cm} \times 16\text{ cm} \times 8\text{ cm} = 1{,}536\text{ cm}^3$. The volume of the tallest part: $10\text{ cm} \times 20\text{ cm} \times 8\text{ cm} = 1{,}600\text{ cm}^3$.

   The volume of the third part: $30\text{ cm} \times 10\text{ cm} \times 8\text{ cm} = 2{,}400\text{ cm}^3$.

   The total volume is: $1{,}536\text{ cm}^3 + 1{,}600\text{ cm}^3 + 2{,}400\text{ cm}^3 = \underline{5{,}536\text{ cm}^3}$.

5. a. The total volume is $107{,}400\text{ in}^3$.

   Part on the left: $V = 38\text{ in} \times 30\text{ in} \times 50\text{ in} = 57{,}000\text{ in}^3$.

   Part on the right: $V = 40\text{ in} \times 30\text{ in} \times 42\text{ in} = 50{,}400\text{ in}^3$.

   The total volume is $107{,}400\text{ in}^3$.

   b. 62.153 ft^3.

Puzzle corner. The edge of the larger cube is 10 in. Therefore, the edge of the smaller cube is 5 in, and its volume is $5\text{ in} \times 5\text{ in} \times 5\text{ in} = 125$ cubic inches. The total volume of the two boxes is then 1,125 cubic inches.

Area and Perimeter Problems, pp. 85-89

1. a. 19 ½ sq. in. The area of the larger rectangle is 10 in × 15 in = 150 sq. in.
   The sides of the white rectangle are 15 − 9 ½ − 2 ¼ = 3 ¼ in and 10 − 2 − 2 = 6 in.
   The area of the white rectangle is thus (3 ¼ in) × 6 in = 19 ½ sq. in.
   b. The shaded area is 150 sq. in − 19 ½ sq. in = 130 ½ sq. in.

2. 324 cm^2. The total area of picture and frame is 38 cm × 29.5 cm = 1,121 cm^2. Subtract the width of the frame (twice, once for each side) from these dimensions to find the width and length of the picture inside the frame:
   38 cm − 2.6 cm − 2.6 cm = 32.8 cm, and 29.5 cm − 2.6 cm − 2.6 cm = 24.3 cm.
   The area of the picture inside the frame is thus 32.8 cm × 24.3 cm = 797.04 cm^2 ≈ 797 cm^2.
   The area of the frame is the difference between the total area and the area of the picture: 1,121 cm^2 − 797 cm^2 = 324 cm^2.

3. a. (1/2 mi) ÷ 4 = 1/8 mi.
   b. To convert from miles to feet, multiply by 5,280: (1/8) × 5,280 = 5,280 / 8 = 660 ft.
   You can also solve this by converting the perimeter of 1/2 mile into feet, and dividing that by 4.

4. a. Student sketches may vary, but should show a small rectangle inside a large rectangle, and the dimensions marked.

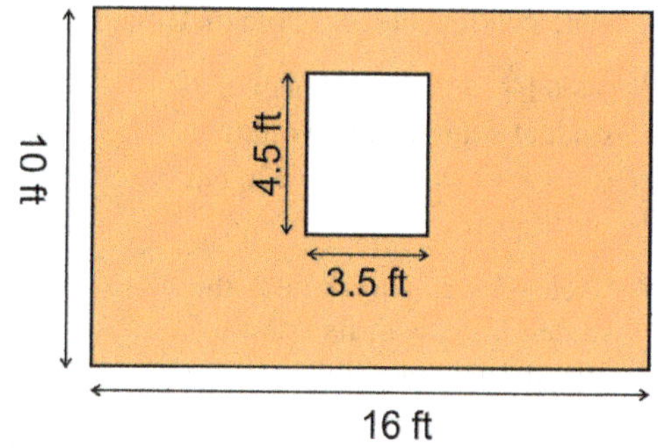

   b. The area of the window is 3.5 ft × 4.5 ft = 15.75 ft^2.

   c. The area of the actual wall is 16 ft × 10 ft − 15.75 ft^2
   = 160 ft^2 − 15.75 ft^2 = 144.25 ft^2.

   d. Since 1 gallon covers 350 sq. ft., it is obviously enough to cover the whole wall.

   e. 1 quart will cover 1/4 of 350 ft^2, which is 87.5 ft^2. Then, two quarts will cover double that, or 175 ft^2. So, two quarts is enough.

5. Rectangle 1 has an area of 4 cm × 4 cm = 16 cm^2.
   Rectangle 2 has an area of 3 cm × 4 cm = 12 cm^2.
   Rectangle 3 has an area of 2 cm × 5 cm = 10 cm^2.
   The total area is: 16 cm^2 + 12 cm^2 + 10 cm^2 = 38 cm^2.

6. a. Area: Divide the figure into five rectangles. The middle rectangle has an area of 6 ft × 6 ft = 36 sq. ft. Each of the outer rectangles has an area of 6 ft × 4 ft = 24 sq. ft. In total, the area is then 4 × 24 sq. ft + 36 sq. ft = 132 sq. ft.
   b. Perimeter: 4 × 6 ft + 8 × 4 ft = 24 ft + 32 ft = 56 ft.

7. The areas of the four component rectangles are: 6 × 12 = 72 cm^2; 9 × 6 = 54 cm^2; 6 × 12 = 72 cm^2; and 6 × 9 = 54 cm^2.
   So, the total area is 72 cm^2 + 54 cm^2 + 72 cm^2 + 54 cm^2 = 252 cm^2.
   The perimeter is 3 + 3 + 6 + 6 + 9 + 6 + 6 + 9 + 6 + 3 + 6 + 6 + 9 + 6 + 6 = 90 cm.

8. The areas of the rectangles 1, 2, 3, and 4 illustrated at right are:
   Rectangle 1: (1 + 1) × (1 + ½) = 2 in × 1 ½ in = 3 sq. in.
   Rectangle 2: (¾ + 1 + 1) × 2 = 2 ¾ in × 2 in = 5 ½ sq. in.
   Rectangle 3: (1 + ½) × ½ = 1 ½ in × ½ in = ¾ sq. in.
   Rectangle 4: 1 × (½ + 2) = 1 in × 2 ½ in = 2 ½ sq. in.

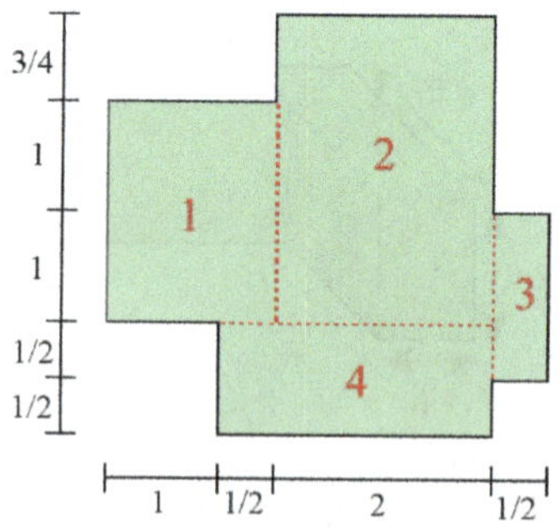

   The total area is the sum:
   3 sq. in + 5 ½ sq. in + ¾ sq. in + 2 ½ sq. in = 11 ¾ sq. in.

   The perimeter is:
   1 ½ + ¾ + 2 + 1 ¾ + ½ + 1 ½ + ½ + ½ + 2 ½ + 1 + 1 + 2 = 15 ½ in.

Star Polygons, pp. 90-91

1. 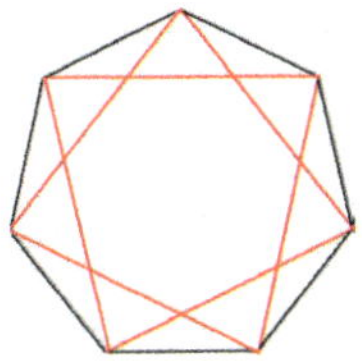

2. a. 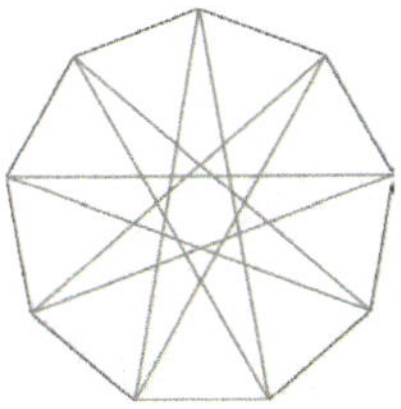 b. 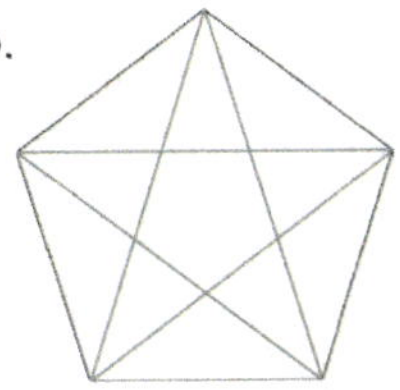

2. c. 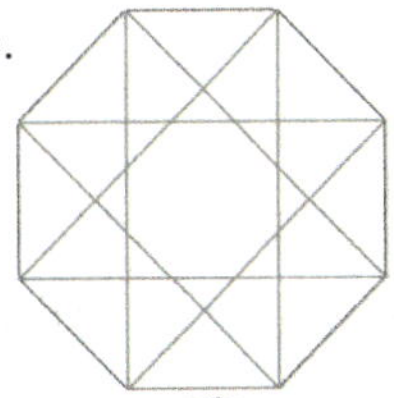 d. 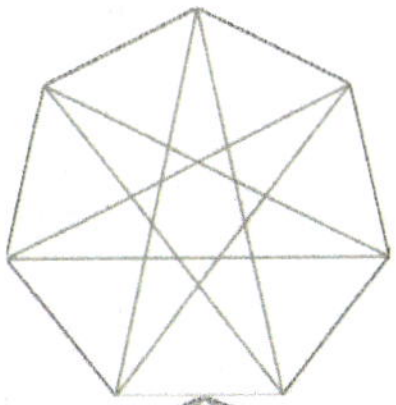

e. 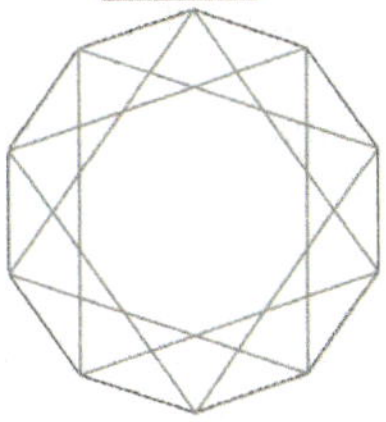 f. 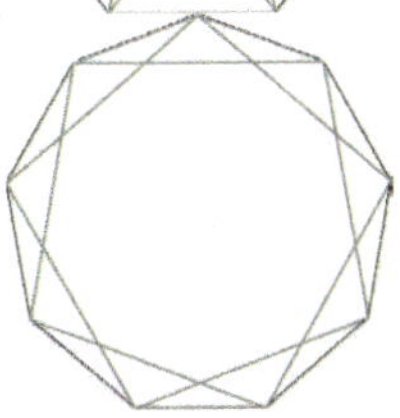

Review 1, pp. 92-95

1. Side 1: 17 1/2 ft; Side 2: 10 ft; Side 3: 17 1/2 ft

2. The area is 320 m^2. First, divide the shape into two rectangles. Use subtraction to find some missing side lengths.

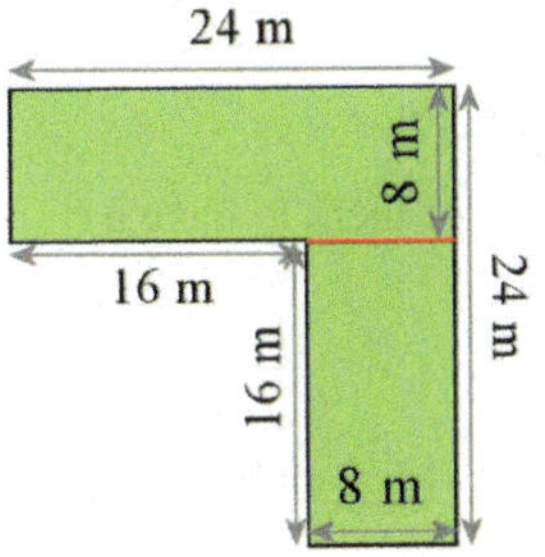

The upper rectangle is 24 m × 8 m so its area is 192 m^2.
The lower rectangle is 8 m × 16 m so its area is 128 m^2.
In total, the area is <u>320 m^2</u>.

3. a. 100° b. 33°

4. a. right b. acute c. obtuse d. acute e. acute
f. right g. right h. obtuse i. right j. obtuse
k. obtuse

5. Check the student's answer.
Here is one such angle:

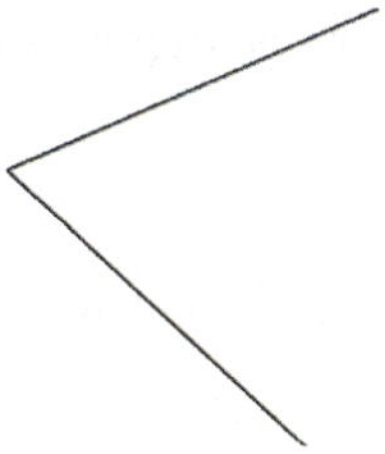

6. It is 84°. The student can figure it out in any manner.
If we use an equation, we write:
$64° + x + 32° = 180°$; $x = 84°$.

7. 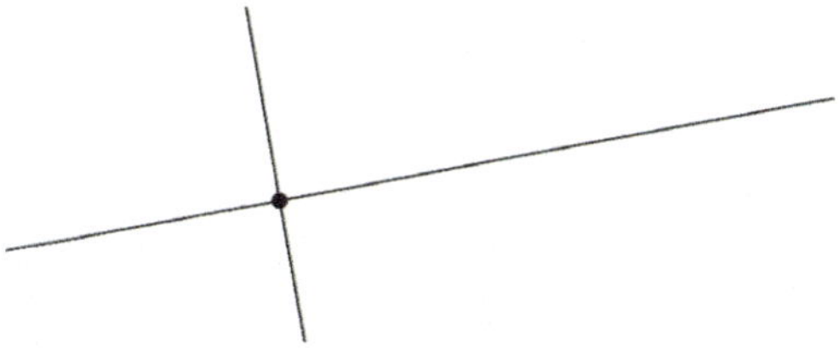

8. Task b. is possible to do. Answers will vary as the other two angles can vary. For example:

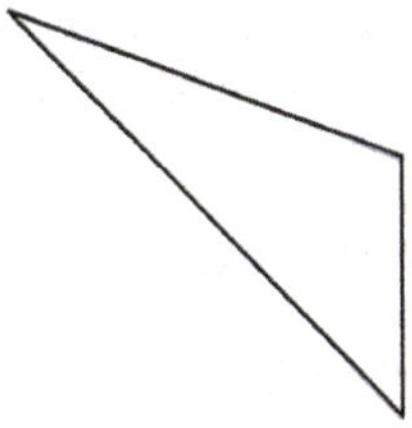

9. Check the student's drawing.

10. They are right triangles.

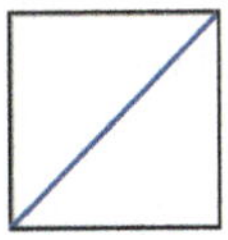

11. $\overline{AB} \parallel m$, $\overline{BC} \parallel n$,
$\overline{AC} \perp m$, $\overline{AB} \perp \overline{AC}$

12. Answers will vary. Check the student's drawing.

13. a. 

b. not symmetrical

c. 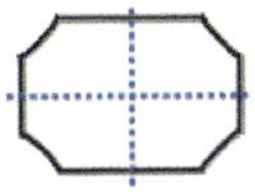

d. not symmetrical

e. 

Review 2, pp. 96-98

1. a. pentagon b. regular octagon

2. a. a regular hexagon b. 120°

3. A kite. Angle A = 139°; Angle B = 53°; Angle C = 139°; Angle D = 29° (or 28)°.

4. a. It is an acute isosceles triangle. b. It is an obtuse scalene triangle.

5. Yes, no, no. An equilateral triangle always has three acute angles (in fact, they are all 60°).

6. Answers will vary. Check the student's answers. For example: a.

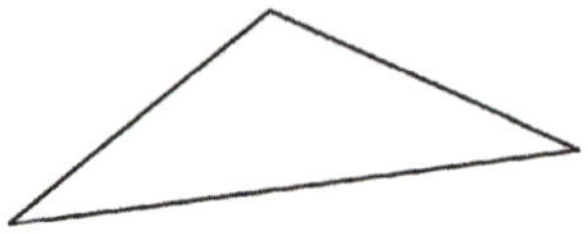

b.

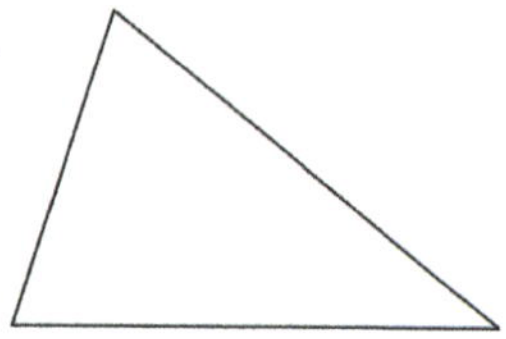

7. a. rhombus b. rectangle c. square d. kite e. parallelogram f. trapezoid

8. a. rectangle b. rhombus c. trapezoid

9. a. The figure has $4 \times 5 \times 2 = 40$ cubes, so its volume is 40 cubic inches.

   b. In this case, each little cube has a volume of 2 in × 2 in × 2 in = 8 cu. in. There are 40 cubes, so their total volume is 40 × 8 cu. in = 320 cu. in. Or, you can calculate the volume by first calculating the three dimensions: the length is 10 inches, the depth is 4 inches, and the height is 8 inches, so the volume is then 10 in × 4 in × 8 in = 320 cu. in.

10. The area of the bottom is 2 cm × 4 cm = 8 cm^2, so the missing dimension is 32 cm^3 ÷ 8 cm^2 = <u>4 cm</u>.

11. Three boxes. One box has a volume of 6 in × 3 in × 2 in = 36 cubic inches. You will need three of them to have a total volume of 108 cubic inches.

12. Starting from the top left corner and going clockwise, the perimeter is:

    $(\frac{3}{4} + \frac{1}{2} + 2) + (1 + 1 + \frac{1}{2} + \frac{1}{2} + \frac{1}{2} + \frac{1}{2}) + (2 + \frac{1}{2} + \frac{1}{2} + \frac{1}{2} + \frac{3}{4}) + 1 + 1 = 3\frac{1}{4} + 4 + 4\frac{1}{4} + 2 =$ <u>$13\frac{1}{2}$ inches</u>.

    The area is: (2 in × $\frac{3}{4}$ in) + (3 in × $2\frac{1}{2}$ in) + ($\frac{1}{2}$ in × $\frac{1}{2}$ in) = $1\frac{1}{2}$ in^2 + $7\frac{1}{2}$ in^2 + $\frac{1}{4}$ in^2 = <u>$9\frac{1}{4}$ in^2</u>.

13. a.

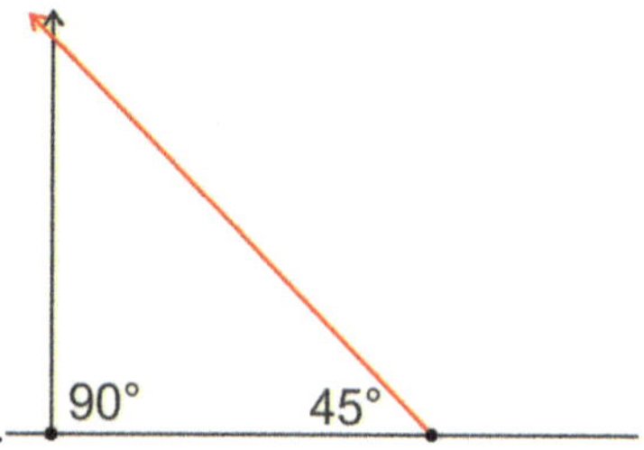

   b. The third angle measures 45°. What the student measures may be slightly different from that, depending on how accurate the drawing is.

   c. The 90° angle is right; the other two are acute.

Puzzle corner. The edge length of the cube must be 4 cm. Therefore, its volume is 4 cm × 4 cm × 4 cm = 64 cm^3.

More from math MAMMOTH

Math Mammoth has a variety of resources to fit your needs. All are available as economical downloads, and most also as printed copies.

- **Math Mammoth Light Blue Series**
  A complete curriculum for grades 1-8. Each grade level includes two student worktexts (A and B), which contain all the instruction and exercises all in the same book, answer keys, chapter tests, end-of-year test, cumulative reviews, and a worksheet maker. International (all metric), Canadian, and South African versions are also available.
  https://www.MathMammoth.com/complete-curriculum
  https://www.MathMammoth.com/international/international
  https://www.MathMammoth.com/canada/
  https://www.MathMammoth.com/south_africa/

- **Math Mammoth Skills Review Workbooks**
  These workbooks are intended to be used alongside the Light Blue series full curriculum, and they provide additional review to the topics studied in the main curriculum, in a spiral manner.
  https://www.MathMammoth.com/skills_review_workbooks/

- **Math Mammoth Blue Series**
  Blue Series books are topical worktexts for grades 1-8, containing both instruction and exercises. They cover all elementary math topics from 1st through 8th grade, from basic addition and multiplication to fractions, percentages, and equations. These books are not tied to grade levels, and are thus great for filling in gaps.
  https://www.MathMammoth.com/blue-series

- **Make It Real Learning**
  These activity workbooks concentrate on answering the question, "Where is math used in real life?" The series includes various workbooks for grades 3-12.
  https://www.MathMammoth.com/worksheets/mirl/

- **Review Workbooks**
  Workbooks for grades 1-8 that provide a comprehensive review of one grade level of math—for example, for review during school break or summer vacation.
  https://www.MathMammoth.com/review_workbooks/

Free gift!

- Receive over 350 free sample pages and worksheets from my books, plus other freebies:
  https://www.MathMammoth.com/worksheets/free

Lastly...

- Inspire4 is an inspirational website for the whole family I've been privileged to help with:
  https://www.inspire4.com

Made in the USA
Monee, IL
14 September 2026

63245909R00066